T0131057

PLATO'S
UNIVERSE

PLATO'S UNIVERSE

Gregory Vlastos

With a new Introduction by
Luc Brisson

PARMENIDES
PUBLISHING

Originally published in 1975
by the University of Washington Press

This paperback edition, with a new Introduction by Luc Brisson,
published in 2005 by Parmenides Publishing
in the United States of America

ISBN-10: 1-930972-13-X
ISBN-13: 978-1-930972-13-1

Printed In Canada

Library of Congress Cataloging-in-Publication Data

Vlastos, Gregory.
  Plato's universe / Gregory Vlastos ; with a new introduction by
Luc Brisson.
    p. cm.
  Originally published: Seattle : University of Washington Press,
1975, in series: The Jessie and John Danz lectures. With new
introd.
  Includes bibliographical references and indexes.
  ISBN-13: 978-1-930972-13-1 (pbk. : alk. paper)
  ISBN-10: 1-930972-13-X (pbk. : alk. paper)
  1. Cosmology—History. 2. Philosophy, Ancient. 3. Plato. I. Title.
  B187.C7V55 2005
  113—dc22   2005032892

Figures 2, 3, 4, and 6 are taken from Figures 1–4 in *Plato*, by Paul
Friedländer, translated by Hans Meyerhoff (Bollingen Series LIX),
vol. 1, *An Introduction* (copyright © 1958 by Bollingen Foundation
and © 1969 by Princeton University Press), and are reprinted by
permission of Princeton University Press.

1-888-PARMENIDES
www.parmenides.com

I dedicate this little book in affectionate gratitude to Princeton's trustees and loyal alumni. On their faith in their university rests its continuing excellence.

The Jessie and John Danz Lectures

(The Series in which *Plato's Universe* was originally published)

In October, 1961, Mr. John Danz, a Seattle Pioneer, and his wife, Jessie Danz, made a substantial gift to the University of Washington to establish a perpetual fund to provide income to be used to bring to the University of Washington each year " . . . distinguished scholars of national and international reputation who have concerned themselves with the impact of science and philosophy on man's perception of a rational universe." The fund established by Mr. and Mrs. Danz is now known as the Jessie and John Danz Fund, and the scholars brought to the University under its provisions are known as Jessie and John Danz Lecturers or Professors.

Mr. Danz wisely left to the Board of Regents of the University of Washington the identification of the special fields in science, philosophy, and other disciplines in which lectureships may be established. His major concern and interest were that the fund would enable the University of Washington to bring to the campus some of the truly great scholars and thinkers of the world.

Mr. Danz authorized the Regents to expend a portion of the income from the fund to purchase special collections of books, documents, and other scholarly materials needed to reinforce the effectiveness of the extraordinary lectureships and professorships. The terms of the gift also provided for the publication and dissemination, when this seems appropriate, of the lectures given by the Jessie and John Danz Lecturers.

Through this book, therefore, another Jessie and John Danz Lecturer speaks to the people and scholars of the world, as he has spoken to his audiences at the University of Washington and in the Pacific Northwest community.

Contents

INTRODUCTION TO THE NEW EDITION
  BY LUC BRISSON   xi

ACKNOWLEDGMENTS   xxi

INTRODUCTION   xxiii

1. THE GREEKS DISCOVER THE COSMOS   3

2. PLATO'S COSMOS, I: THEORY OF CELESTIAL
  MOTIONS   23

3. PLATO'S COSMOS, II: THEORY OF THE
  STRUCTURE OF MATTER   66

APPENDIX   98

BIBLIOGRAPHY   117

INDEX OF GREEK WORDS   123

INDEX OF NAMES   125

INDEX OF PASSAGES   127

Introduction to the New Edition

By Luc Brisson

GREGORY Vlastos was Turkish by birth, Greek by blood, and American by immigration. He was born on July 27, 1907, in Istanbul and raised in the Protestant faith. His undergraduate work was performed at Istanbul's Robert College, an American-funded institution. After his graduation in 1925, Vlastos pursued studies in theology at the University of Chicago, where he obtained a bachelor's degree in theology in 1929; he was subsequently ordained a minister. Vlastos then went on to study philosophy at Harvard University, where he obtained his Ph.D. with a thesis titled *God as a Metaphysical Concept*, written under the direction of Alfred North Whitehead.

Between 1931 and 1948, Vlastos resided in Canada, where he taught at Queen's University, Ontario, and became a Canadian citizen. In 1938, he left for Cambridge to work with Francis MacDonald Cornford. In 1939, with Cornford's support, Vlastos published an article titled "The Disorderly Motion in the *Timaeus*," in which he criticized the position of the author of a famous commentary on the *Timaeus* (1937) on the question of whether the origin of the world should be situated within time. When World War II broke out, Vlastos joined the Royal Canadian Air Force, serving as a squadron leader in the Personnel Division and as editor of *Canadian Affairs*, a publication of the Wartime Information Board. At the end of the war, Vlastos returned to academia and published "Ethics and Physics in Democritus" (1945) and "Equality and Justice in Early Greek Philosophy" (1947).

Vlastos left Canada in 1948 for a position at the Sage School of Philosophy at Cornell University, a center of the new style of analytic philosophy imported from Britain and Austria, which was taking over American philosophy. There, he was initiated into the new methods by Max Black and was colleagues with Friedrich Solmsen. These years saw the publication of Vlastos' "The Physical Theory of Anaxagoras" (1950), "Theology and Philosophy in Early Greek Thought" (1952), and "Isonomia" (1953).

Vlastos was invited to be a fellow of the Institute for Advanced Study at Princeton University in 1954–55. There he was able to work with another leading student of ancient philosophy, Harold Cherniss. Cherniss's seminar on Heraclitus influenced Vlastos' own study of that philosopher, as evidenced in Vlastos' "On Heraclitus" (1955). Vlastos then published his famous article "The Third Man Argument in the *Parmenides*" (1954), which was to unleash a flood of papers following replies by two leading philosophers, Wilfred Sellars and Peter Geach.

In 1955, Vlastos was invited to join the philosophy department at Princeton as a Stuart Professor. He became an American citizen in 1972, which was the same year he delivered the Jessie and John Danz lectures at the University of Washington, published in 1975 under the title *Plato's Universe*. These lectures were followed in 1973 by the publication of *Platonic Studies*, which contains the majority of his most important articles published until that time. In the same year, students and colleagues honored him with a Festschrift called *Exegesis and Argument: Studies in Greek Philosophy*, presented to Gregory Vlastos and edited by E. N. Lee, A. P. D. Mourelatos, and R. M. Rorty.

Vlastos formally retired from Princeton in 1976 and moved to Berkeley, California, where he became a permanent Mills Visiting Professor, teaching seminars to graduate students and, on seven occasions, his National Endowment for Humanities Summer Seminar on the Philosophy of Socrates. These courses, together with several series of lectures, gave rise to the publication in 1991 of his *Socrates, Ironist and Moral Philosopher*. Vlastos passed away on October 12, 1991. His articles were collected and published under the titles *Socratic Studies* (1994) and *Studies in Greek Philosophy* (1994).

Vlastos' theological training and political commitments led his writings to encompass more than just the field of Greek philosophy; his concerns about political philosophy and the religious and ethical foundations of democracy are often perceptible in his scholarly works. In addition, the development of his career as a researcher and a teacher explains why one finds in Vlastos' works on Greek philosophy a happy combination between the analytic approach (practiced primarily in England and in North America), which is exclusively interested in the structure of arguments, and the hermeneutic approach (practiced in continental Europe), which is concerned with the historical context and takes religious and ethical questions into consideration. Moreover, this is why the conference presented to Vlastos was titled *Exegesis and Argument*.

At the same time, it seems to me that his theological training and his interest in the hermeneutic approach were motivating factors for Vlastos' explorations of the *Timaeus,* a dialogue that evokes the origin of the world and describes its constitution. In the Anglo-Saxon world after World War II, the analytic approach was established and consolidated: it cut ancient philosophy off from its concerns for history in order to anchor it within the analysis of ordinary language and argumentation. This retreat toward linguistics and logic placed Aristotle in a superior position over Plato and led to the acceptance of Aristotelian criticisms of Plato. In practice, this position implied shelving Platonic metaphysics; hence, one refused to speak of the soul and of Forms in Plato. The most significant article in this regard remains G. E. L. Owen's "The Place of the *Timaeus* in Plato's Dialogues" (1953), which continues to exert some influence. It sought to show that the *Timaeus,* with its doctrine of Forms, belonged to a previous period of Plato's thought, that of the *Republic,* which was also that of the "mad Plato" who still believed in the reality of those phantasms that Kant definitively rejected as outside the domain of objective knowledge. In the *Parmenides,* Plato supposedly questioned the doctrine of the Forms, which he henceforth considered as concepts, as can be seen in the *Sophist.* Cherniss replied to these arguments in a well-known article titled "The Relation of the *Timaeus* to Plato's Later Dialogues" (1957). However, in the Anglo-Saxon philosophical world, the *Timaeus* continued to be considered a dialogue

sui generis to which only very few works were devoted. Vlastos'
Plato's Universe, which takes up several works and leads them
to their conclusions, constitutes the first attempt to break
through this isolation, hence its importance for Plato's image
in the English-speaking world.

Recalling that many important things have come to us from
the Greeks—democracy, tragedy, the Olympic Games, mathematics, logic, philosophy—Vlastos wonders in the introduction to *Plato's Universe* if the Greeks really discovered what
we now mean by "science." Less optimistic than Sambursky
and Burnet, Vlastos restates the reserved answer he had already given in 1955, saying that even if they were not able to
"grasp the essential genius of the scientific method," they did
discover the notion of a cosmos "that is presupposed by the
idea of natural science and by its practice." In fact, the early
Greeks had "the perception of a rational universe." This is
why the first chapter of this book is devoted to the 6th and
5th centuries B.C.—that is, to the *physiologoi*—and is titled
"The Greeks Discover the Cosmos." After giving the back-
ground of the *Timaeus,* Vlastos turns his attention to "Plato's
role in the reception and transmission of the discovery of
the cosmos." Plato enabled the radical transformation of this
notion of the cosmos by annexing it to his "idealistic and
theistic metaphysics." Very different opinions have been
voiced on the effects of this transformation. Some scholars
have denounced Plato's obscurantist and reactionary attitude,
which allegedly blocked the development of science; others
have maintained that Plato's approach was in accord with that
of modern scientists ranging from Galileo to Heisenberg.
However, having previously maintained the latter, in this book
Vlastos adopts a "middle-of-the-road" position. Whereas in
his second chapter, "Plato's Cosmos I: Theory of Celestial
Motions," Vlastos describes the positive aspects of Plato's approach, in his third chapter, "Plato's Cosmos II: Theory of
the Structure of Matter," he points out its shortcomings.

Generally speaking, the ancient Greeks accepted the following definition of *physis* (nature): "the *physis* of any given
thing is that cluster of stable characteristics by which we can
recognize that thing and can anticipate the things within
which it can act upon other things or be acted upon them."

However, they also considered that "*physis* fixes the limits of the possible for everything except the supernatural." The gods, who in heaven and on earth were motivated by a combination of competitiveness (*agon*) among themselves and with regard to men, did not, when pushed by jealousy (*phthonos*), deprive themselves of the pleasure of interfering in the workings of the universe by causing eclipses and disasters or by modifying the mental processes of human beings to make them act against their will.

All this changed with the *physiologoi*, the thinkers of the 6th and 5th centuries B.C. who inquired into the origin of the world in which we live. There were no longer any exceptions to natural processes, no supernatural interventions. The positions of these thinkers could be diverse, with Heraclitus and Democritus being the most demanding. Yet they all accepted the following position: "For the first time in history man has achieved a perception of a rational universe which leaves his own destiny to be determined solely by *physis*—his own and that of the world." In short, while one cannot say that a scientific explanation of the world had been given, it must be admitted that we find the certainty that the behavior of the world and of man is rational, without any divine interference.

In Book X of his *Laws*, Plato decrees a law on impiety, the equivalent of which cannot be found elsewhere. Plato accuses the poets and *physiologoi* of being responsible for their compatriots' impiety. He reproaches the *physiologoi* with having explained the universe by purely mechanical causes. Yet what does he have to oppose them? Vlastos' answer is clear and simple: a "theological cosmogony." "Plato undertakes to depict the origin of the cosmos as the work of a god who takes over matter in a chaotic state and moulds it in the likeness of an ideal model," Vlastos states. Does this god have the power to modify the regularities of nature? One cannot say. What is clear, however, is that even if he has such power, he will never exercise it, for the demiurge features two essential characteristics when compared to traditional divinities: not only is he reason personified, but he is above all bereft of jealousy (*phthonos*). This is why he is "driven by the desire to share his excellence with others." In short, although Plato accepts the idea of divinity, all possibility of supernatural intervention in the regularities of nature is impossible. It is from this perspective that Vlastos proposes to read Plato's *Timaeus*. In the first part, he turns to the triumph of rationality (described in *Timaeus* 29e–47d); then, in the second part, he evokes the compromises rationality must enter into with necessity (described in *Timaeus* 47e, near the end).

Vlastos distinguishes two sets of propositions in this first section, some of which deal with metaphysics and theology, others with science. The presuppositions touching upon metaphysics and theology are as follows: the cosmos has a soul; it is unique; it has a spherically formed body; and it is located within time. These presuppositions play a role in scientific explanations, particularly in the field of astronomy. Since the cosmos has a soul, the demiurge starts by occupying himself with the fashioning of this soul. As an intermediate between the sensible and the intelligible, the soul, which is the self-moving principle of all motion in the world, must first account for the circular motions of the celestial bodies, with the circle associated with one of the three types of mathematical relationships that present the closest earthly image of the regularity and invariance that characterize the Forms. Henceforth, Plato is in a position to propose an astronomical model based on the following four theses: 1) the stars are gods and their motions are psychokinetic; 2) stellar motions are circular; 3) the souls of the star-gods are perfectly rational; and 4) all rational motion is circular. These theses are associated with the hypothesis that the composition of postulated regular motions may account for irregular phenomenal motions.

This was a highly fruitful hypothesis that enabled the development of astronomy. A conceptual matrix that generates information such as I have just detailed leaves no doubt as to its scientific value, even if Plato obtained it by way of his metaphysical scheme. This can be explained by the fact that Plato was aware of "scientifically ascertained facts," whether through his own resources or through the intermediary of members of the Academy—for instance, Eudoxus. These facts are defined by three features: 1) they are established by observation or by inference from it and are derived, directly or indirectly, from the use of the senses; 2) they have theoretical significance, providing answers to questions posed by theory; and 3) they are shared and corrigible, the common property of qualified investigators who are aware of possible sources of observational error and are in a position to repeat or vary the observation. To be sure, Plato reintroduces supernatural forces into his cosmos, but these forces cannot intervene against the regularity of phenomena and even constitute an absolute guarantee of this regularity. Paradoxically, metaphysics and theology favor the development of a genuine scientific explanation. However, this scientific explanation had its limits: it was to license Plato to rest content with a purely kinematic model that aimed to show how, if certain motions were assumed, the mathematically deduced consequences would save the phenomena.

Vlastos then moves on to the second chapter on the *Timaeus*. When we proceed from the first division of the *Timaeus* (29e–47e), which deals mainly with the teleologically ordered motions of souls, to the second division (47e–69b), which deals with the mechanistically ordered motions of earth, water, air, and fire, the evidence that Plato shares the outlook of his adversaries becomes far more compelling. Continuing along the lines of the *physiologoi,* whom he criticizes elsewhere, Plato admits that the unobservables (the corpuscles) we postulate to account for the properties of the observables (the sensible particulars) need not themselves possess those same properties. The material that confronts the demiurge in its primordial state is inchoate. The four primary kinds of matter—earth, water, air, and fire—are present here as a blurred, indefinite form, and their motion is disorderly. The demiurge changes all this. He transforms matter from chaos to cosmos by impressing on it regular stereometric form. The corpuscles formed in this way undergo two types of transformation: one concerns the corpuscles themselves; the other concerns the varieties of each of the elements. Unlike what occurs in the field of astronomy, in the field of physics Plato could not count on the observation of scientifically ascertained facts. Hence, it became impossible to give priority to one model over another. How, then, can we explain this absence of confrontation with empirical facts? Vlastos finds the answer in the *Timaeus* itself (68c–d): only a god could achieve such a confrontation, and man is not god. Consequently, in the *Timaeus* Plato offers only something believable, something that *could* be true.

Three critical remarks can be made today with regard to this essential work. The first concerns the fact that Vlastos remains dependent on the ideas of his time about the relations between philosophy and religion in both the *physiologoi* and Plato, two fields that never cease to interpenetrate each other and are never really opposed. This is particularly clear with regard to the fragments of Heraclitus that evoke the *Logos* and the origin of the world in the *Timaeus*. In addition, it is not at all certain that the new theories of matter would prove Democritus right and Plato wrong, especially insofar as the constitution of a mathematical model is concerned.

The second concerns the relations of Plato's *Timaeus* to Aristotle. I shall insist on only two points. When he mentions astronomy, Vlastos speaks of "teleology," a term that is never found in Plato and that results from the criticism Aristotle makes of Plato on the basis of his theory of the four causes: material, formal, efficient, and final. Moreover, every time Vlastos speaks of teleology, he refers to passages where Plato speaks of rationality. The problem is even more complex when we come to physics in the proper sense of the term. Here, Vlastos speaks of "matter," a term that is never found in Plato but which is used by Aristotle as a synonym for the term *khora*. This is crucial, for in the *Physics* Aristotle develops a very advanced critique of the *khora* mentioned in the *Timaeus,* assimilating it to space and thus criticizing the representation of the elements as regular solids. In fact, Aristotle wants to show by this that it is absurd to wish to give the concrete physical world (where resistance reigns) an abstract explanation (in mathematical terms). Here we reach the most important point of disagreement between Plato and Aristotle. Aristotle wants to provide an explanation of nature that is based on ordinary language, whereas Plato wants to leave such an explanation up to mathematics, which, in ever-changing sensible things, represents the traces of intelligible stability. In fact, since the limits of Plato's physical explanation coincide with the limits of the mathematics of his time, Aristotle could show himself to be dissatisfied with the results obtained, yet his opposition in principle is untenable.

Finally, Vlastos is perfectly right to hold that the absence of observation and above all of experimental verification removes all scientific pretensions from the physics of the *Timaeus*. Yet the reasons for this absence are more numerous and complex than he believes. First, to be sure, humankind cannot equal god. Second, we must take into consideration the cultural factor that, for an ancient Greek, proof pertained not to experimental verification but to rhetorical proof, that is, illustration. Third, carrying out an experimental verification required technology that the Greeks did not have. And finally, the Greeks of Plato's time had neither a satisfactory numerical system nor a universal system of measurement. All these handicaps made true experimental verification impossible and thus made all pretensions to science illusory.

Properly speaking, this work is not a commentary on Plato's *Timaeus*. For Vlastos, the standard reference in this field remains the book by Cornford. Instead, the book is, in fact, an inquiry into whether or not the ancient Greeks were at the origin of science. Vlastos' answer remains prudent. The ancient Greeks invented the notion of cosmos, which is presupposed by the idea of a science of nature and by its practice. This idea implies that the regularities in nature cannot be challenged by the intervention of divinities either in the world or in humankind. Essentially, I subscribe to this answer, which enables us to have a fine appreciation of Plato's relations with his predecessors, the *physiologoi*, while maintaining an emphasis on the originality of Plato's approach, which reintroduces divinity into the world—albeit a new type of divinity that cannot challenge the regularities of nature.

The renewed availability of Vlastos' book to today's scholars and students is truly important in several respects. The clarity of Vlastos' argumentation, his historical knowledge, and his familiarity with the text enable us to make contact once again with one of Plato's works in which little interest has been shown in the English-speaking world since the Second World War: the *Timaeus*. They also enable us to see that in this work, Plato pursued in an original way the research on nature and the cosmos that had been inaugurated in Greece a few centuries before him. What is more, the life and work of Vlastos constitute a particularly interesting testimony to recent history, as far as the links between religious convictions and positivistic thought, and hermeneutics and analytic philosophy are concerned. All these aspects make this new edition of *Plato's Universe* a most welcome addition to both Platonic studies and the history of 20th-century scholarship.

Luc Brisson, CNRS
Paris, October 2005
(trans. Michael Chase)

Acknowledgments

It is a pleasure to acknowledge the hospitality I received on the campus of the University of Washington from Professor David Keyt, chairman of the Department of Philosophy, and Professor John McDiarmid, chairman of the Department of Classics, and from many of their colleagues, during the summer of 1972, when I delivered these lectures; and the timely help provided by Edward Johnson who prepared, on very short notice, the three Indexes of the book.

G. V.

Introduction

SUCH a variety of wonderful things have come to us from the Greeks: democracy, tragedy, Olympic games, the nude, pure mathematics, formal logic, philosophy. Is our idea of science also a Greek heritage? Did they really discover what we now mean by "science"? In some of our sciences they made brilliant advances—in astronomy, for example, and scientific history. But it is one thing to learn how to do something and do it well, quite another to understand what it is that you are doing—to conceptualize its methods, to formulate its principles. Distinguished historians have not hesitated to claim just that for the Greeks. Sambursky in his *Physical World of the Greeks* (1956) declares roundly that "the basic principles of the scientific approach . . . were discovered in Ancient Greece."[1] He locates the discovery in the sixth century B.C., assigning it to the succession of thinkers who had been called *physiologoi*, "men who discoursed on nature." The most famous statement of this claim is in the Preface to the Third Edition of John Burnet's classical treatise, *Early Greek Philosophy* (1920): "My aim has been to show that a new thing

---

1. P. 3. Again, on p. 52 he speaks of "the new, intensely rational, scientific method introduced by the Milesian School in the very early days of Greek science." But before he gets through, Sambursky provides some good correctives. See the penetrating analysis of the "Limits of Greek Science" in the final chapter, especially the remarks on the failure of the Greeks to achieve in the domain of terrestrial physics the "dissection of nature" by experiment (pp. 23ff.).

came into the world with the early Ionian teachers—the thing
we call science—and that they first pointed the way which Europe has followed ever since, so that . . . it is an adequate
description of science to say that it is 'thinking about the
world in the Greek way.'"

I contested that claim nearly twenty years ago and gave
briefly my reasons.[2] I shall have more to say on that score
here, especially in the third chapter. But I do not wish to
make too much of this negative thesis. Far more interesting
for us, and more important for humanity than the failure of
those early Greek thinkers, is their achievement. Though it
was not given to them—nor, for that matter, to Plato or to
Aristotle after them—to grasp the essential genius of the scientific method, they did discover something else which may
still be reckoned one of the triumphs of the rational imagination: the conception of the cosmos that is *presupposed* by the
idea of natural science and by its practice. This is their contribution to what is termed in the John Danz trust "man's perception of a rational universe," and this is what I shall be
talking about in the opening chapter. I want to recapture the
original sense of the word *kosmos*[3] and try to explain how it
happened that a word with just that sense came to epitomize
the intellectual revolution that began with the cosmogonies
of Thales, Anaximander, and Anaximenes in the sixth century B.C. and culminated a century later in the atomic system
of Leucippus and Democritus. When these men glimpsed
with wild surmise on their Ionian Dariens a new physical universe, why did they pick *kosmos* to name what they saw? What
were they fighting against, and what were they fighting for?
These are the questions I shall be trying to answer in the first
chapter.

In the next two I shall move from the sixth and fifth centuries into the fourth to discuss in some detail Plato's role in

---

2. Vlastos, review of F. M. Cornford, *Principium Sapientiae*, in D. J. Furley
and R. E. Allen, eds., *Studies in Presocratic Philosophy*, 1:42–45 (corrected reprint of the review-article which appeared originally in *Gnomon*, 1955).

3. I write *"kosmos"* when transliterating the Greek word, otherwise "cosmos."

the reception and transmission of the discovery of the cosmos. Why Plato? Why give him the lion's share? My reasons have nothing to do with that absurdly inflated estimate of his place in European thought which is conveyed in Whitehead's oft-parroted remark that the whole subsequent history of Western philosophy consists of footnotes to Plato. We do not need to pay Plato such extravagant compliments to be assured of his greatness as a thinker and writer and to feel the peculiar interest of his own responses to the major intellectual currents of his age. Coming at the end of the movement which first projected and then consolidated the new conception of the universe, he gives us the chance to see that movement from a unique and enlightening slant: that of a fierce opponent of the revolution who wrested from it its brilliant discovery, annexed its cosmos, and refashioned it on the pattern of his own idealistic and theistic metaphysics.

The effect of that transformation has been variously assessed. Some scholars have denounced it as a reactionary, obscurantist move which would have blocked the subsequent growth of natural science, had it prevailed. Others have extolled it, claiming that Plato's conception of the physical world would be more congenial to the creative scientist of the modern era from Galileo to Heisenberg than would be that of Democritus.[4] Thirty years ago I would have sided strongly with the former group. I took that line briefly in a paper published in 1941.[5] What I have since learned from living with Plato's work has shown me that both that view and its opposite are half-truths, or less—that the truth, viewed fairly and dispassionately, lies somewhere in between. The reader may, or may not, come to agree with me. At any rate, I shall try to make it easier for him to reach a balanced view by presenting the best and the worst in Plato's theory of the cosmos.

4. For good samples of opinions on both sides, see the references in G. E. R. Lloyd, "Plato as a Natural Scientist," p. 78.
5. "Slavery in Plato's Thought," *Platonic Studies*, pp. 147–63 (corrected reprint of the essay which appeared originally in the *Philosophical Review*, 1941).

Plato's Universe

1

The Greeks
Discover the Cosmos

In English *cosmos* is a linguistic orphan, a noun without a parent verb. Not so in Greek which has the active, transitive verb, *kosmeō*: to set in order, to marshal, to arrange. It is what the military commander does when he arrays men and horses for battle; what a civic official does in preserving the lawful order of a state; what a cook does in putting foodstuffs together to make an appetizing meal; what Odysseus' servants have to do to clean up the gruesome mess in the palace after the massacre of the suitors.[1] What we get in all of these cases is not just any sort of arranging, but one that strikes the eye or the mind as pleasingly fitting: as setting, or keeping, or putting back, things in their proper order. There is a marked aesthetic component here, which leads to a derivative use of *kosmos* to mean not *order* as such, but *ornament, adornment*; this survives in the English derivative, *cosmetic*, which, I dare say, no one, without knowledge of Greek, would recognize as a blood-relation of *cosmic*. In the Greek the affinity with the primary sense is perspicuous since what *kosmos* denotes is a crafted, composed, beauty-enhancing order. Now for the Greeks the moral sense merges with the aesthetic: they commonly say *kalos*, "beautiful," or *aischros*, "ugly," to mean *morally admirable* or *repugnant*. We would then expect moral, not less than military, civic, domestic, and architectural ap-

---

1. For illustrative texts and illuminating remarks on the meaning of *kosmeō*, see Charles Kahn, *Anaximander and the Origins of Greek Cosmology*, pp. 219ff.

plications of *kosmos*. And this is what we find. In Homer we see it used adverbially (κόσμῳ κατὰ κόσμον) for speech or action conducted in a socially decent, morally proper way.[2] Later we see it even used to signify the general observance of morality and justice, as when Theognis in mid-sixth century, lamenting the corruption of morals, exclaims,

> They seize property by violence; *kosmos* has perished;
> Equitable distribution no longer obtains.
>
> [Vv. 677–78]

What then could have led men who had used *kosmos* in these ways since their childhood to give it a physical application—to use it as a name for a physical system composed of earth, moon, sun, stars, and everything in or on or between these objects, so that if they believed that there was only one such system in existence they would speak of it as "the cosmos" or "this cosmos," while if they thought there were many such they would speak of *kosmoi* in the plural? For the answer I shall look to the fragments of Heraclitus (first third of the fifth century B.C.). These are the earliest surviving sentences in which the word is used with this new sense, and it is sound historical method to work so far as possible with original source-materials. Throughout this part of my discussion it would be well to bear in mind that Heraclitus is not a typical *physiologos*—he is too much of a mystic, a poet, and a metaphysician to fit the general pattern. But though a maverick, he is more, not less, than a *physiologos;* in any case, he is enough of one—he has a theory of the constitution of the physical universe and many explanations of natural phenomena—to make his own use of the term representative.

I begin with fragment B30:[3]

> This cosmos, the same for all, no god or man has made, but it

---

2. Good examples in ibid.

3. The numbering of Heraclitean and Presocratic fragments is that in H. Diels and W. Kranz, *Die Fragmente der Vorsokratiker* (subsequently cited as DK). "B" prefaces the number when the editors judge that the fragment retains the original wording.

was, is, and will be for ever: ever-living fire, kindling according to measure and being extinguished according to measure.[4]

As everyone knows, fire is Heraclitus' universal substance: the whole world, he thinks, consists of fire, though only that part of it which is in the "kindled" state is recognizable as fire and is generally so called. The two other world-masses which must be added to account for the whole of his material universe, water and earth, must then be fire "extinguished": water is fire liquefied, earth is fire solidified. Holding that everything is in constant change—this is his most famous doctrine, the one which makes him for all subsequent ages the philosopher of flux—he takes the world-process to consist of the ceaseless intertransformations of these three components. In each of these interchanges one element assures its own existence by destroying another; in his own language (B76) each "lives the other's death."

To conjure up a physical model that would fall within his own experience, let us take an oil lamp. Its flame exists ("lives") by the constant extinction ("death") of the oil (liquid, hence "water"); so fire "lives the death of water." The same would be true of a wood or charcoal fire, where the victims are solids, and would probably count as "earth" for Heraclitus. So "fire kindling" is water or earth turning into fire, and "fire being extinguished" would be the converse, fire turning into water or earth, thus "living" its own "death." This is happening always, and always "according to measure." If the "measures" of the converse processes, fire kindling and fire extinguished, were the same in all occurrences of fire in the universe, fire would indeed be "ever-living." For then as much of fire would be turning into water and earth at any given time, as of water and of earth into fire at the same time, and then the quantity of fire would remain constant. And if the corresponding thing happened in the

---

4. For a discussion of the text and meaning of this fragment see my paper, "On Heraclitus," pp. 344–47. For fuller discussions see G. S. Kirk, *Heraclitus*, pp. 307–24; W. K. C. Guthrie, *A History of Greek Philosophy* 2: 454–59; Jula Kerschensteiner, "Kosmos," pp. 97–110; M. Marcovich, *Heraclitus: Greek Text with Short Commentary*, pp. 261–71.

case of water and of earth, their quantities too would remain constant. And since these three compose all the matter there is, its distribution as between fire, water, and earth would remain invariant, and the universe as a whole would be eternal, in spite of incessant change throughout its length and breadth.

This would be a striking innovation in natural philosophy. From its beginnings in the sixth century, in Thales, Anaximander, and Anaximenes, its dominant pattern had been cosmogonic: physical theory centered in the question, "What is the source and origin of our world? What is that from which our world began and to which it will eventually return?" The reasoning I have just sketched in Heraclitus would break with this tradition. It would show him how the world could be everlasting, birthless and deathless as a whole, because birth and death keep balancing out within its parts. Here, for the first time in Greek history, we have a cosmology without a cosmogony.

To return to our fragment: "ever-living fire" at the start of the second colon stands in apposition to "this cosmos" at the start of the first.[5] So the two phrases have the same denotation. And since the second refers to the whole world, so must the former. "Cosmos" then cannot mean here simply *order, arrangement*, as many scholars have thought,[6] but that which *has* this order, this arrangement; it must mean the world in its aspect of order.[7] And we can see what kind of order Heraclitus is thinking of here: not the static order of architectonic

5. If we keep the punctuation I have used above (favored by many scholars, including Kirk and Marcovich). Alternatively the colon could be deleted, in which case "ever-living fire" would be the predicate nominative of the verbs in the immediately preceding clause; then Heraclitus would be saying that "this world . . . *is* . . . ever-living fire," but the meaning would not be substantially affected, since in that case the "is" expresses simple identity.

6. Notably Kirk, *Heraclitus*, pp. 307ff., and again in *The Presocratic Philosophers*, p. 199. Rebuttal by Marcovich, *Heraclitus*, pp. 269–70.

7. Cf. my remark in the paper cited in n. 4 above: in this fragment "*kosmos*, though it implies, does not just mean, 'order,' for what is in question here is not merely that nobody made the *order* of the world, but that nobody made this orderly *world*; this world is fire, and nobody made the fire, for it is 'ever-living'" (p. 346).

pattern—no reference, here or elsewhere in Heraclitus, to the shape or structural design of the universe—but to the dynamic order which marks the intertransformations of its elements. In the world picture of his predecessor, Anaximander, spatial, architectonic symmetries have been conspicuous: there sun, moon, stars were a sequence of huge concentric fire-rings around a central earth, with equal intervals between successive pairs in the sequence; and the infinitely many worlds in his universe were at equal intervals from one another.[8] In Heraclitus the symmetries are exclusively temporal: they are causal sequences. The order of his world reveals itself in the constancy of the "measures" of its balanced kindlings and extinguishings.

And if the aesthetic and moral nuances in *kosmos* are recalled, it is not surprising to see Heraclitus refer to the world-order as "Justice," on one hand, as "Harmony," on the other:

> We should know that war is common, and strife is justice and that all things happen according to strife and just necessity. [B80]

> The adverse is concordant; from discord the fairest harmony. [B8]

Fire, water, and earth are in deadly opposition. They are at war. They are annihilating each other. But the dynamic symmetries of their intertransformations harmonize the warring opposites, make them perpetuate each other through their very strife and thus compose a world which is everlasting because all through it life is perpetually renewed by death. If you believe that "from discord [comes] the fairest harmony," what could be fairer than this? And why should it be "just"? Because of the same pattern, now viewed in its aspect of the even-handed reciprocity that insures to each of the three elements its equal share of living and of dying; what each loses in dying into its peers it invariably recoups in living their death.[9]

Now look at the phrase which follows "This cosmos" in

8. For the references see Vlastos, "Equality and Justice in Early Greek Cosmologies" in Furley and Allen, eds., *Studies in Presocratic Philosophy*, 1: 75-76 (corrected reprint of the original in *Classical Philology*, 1947).

9. See my paper cited in n. 4 above, pp. 356-57.

B30: "This cosmos, *the same for all.* . . ." If you wonder, what is the point of saying that, look at the next three fragments:

> For those who are awake the cosmos is one and common. But those who are asleep turn aside each into a private cosmos. [B89][10]
>
> We should not act and speak like men asleep. [B73]
>
> One should follow the common. But while the Logos is common, the many live as though they had a private understanding. [B2]

"Same for all" in B30, "one and common for all" in B89 are obviously equivalent expressions. Now who are those for whom the cosmos is "one and common" in B89? They are men wide awake in contrast to dreamers, to men asleep. Brooding on that bizarre mode of consciousness into which we all sink in our sleep, Heraclitus finds the key to its enfeebled intelligence in its privacy—its anarchic subjectivity. In sleep he sees us cutting ourselves off from the common world which sets the norms of veridical perception, coherent speech, and effective action. That he has both speech and action in view is clear from B73: "We should not act and speak like men asleep." If we talk in our sleep we babble; and the actions we dream we are doing are impotent to achieve their end.[11] So when Heraclitus exhorts his fellows *not* to act and speak like that, he is implying that this is their present condition: they are living in a world as false as that of the dream. And in the next fragment, B2, he laments that "while the Logos is common, the many live as though they had a private understanding." "Logos" here is not only his own discourse but, no less, what his discourse is meant to reveal, that is, the intelligible pattern of the world process. "That is what you must understand and follow," he is telling his contemporaries, "else you will not be able to think, speak, or act

---

10. On the authenticity of this fragment (rejected in whole or in part by some scholars: e.g., Kirk, *Heraclitus*, pp. 63–64, who thinks it a Plutarchean paraphrase) see Kerschensteiner, "Kosmos," pp. 99ff., and Marcovich, *Heraclitus*, pp. 369–70.

11. As in the famous simile in the *Iliad* (bk. 22, vv. 199–200): "As in a dream we cannot run down one who flees;/Powerless he is to escape, and we to pursue."

straight." He is thus claiming that his teaching about the cosmos conveys a truth whose apprehension will change their mode of consciousness, their very lives—that, if they but grasp what he is talking about, their perceptions, their speech, their actions will alter as dramatically as would those of a sleepwalker if he were to be suddenly jarred awake.

What could have prompted him to make such an extraordinary claim? A full answer to that question would take me well beyond the limits of my present theme: it would require me to reckon with Heraclitus' mysticism, that is to say, with that very strain in his thought which makes him untypical of Ionian *physiologia*. So I must content myself with a partial answer which, however, will do for my present purposes, for it takes account of just that part of his outlook which he shares fully with the *physiologoi* and throws good light on their discovery of the cosmos. For this I go to fragment B94:

> The Sun will not overstep his measures, else the
> Furies, the adjutants of Justice, will find him out.

What "measures" could he have had in view here? Each of the following regularities would have qualified:

1. That of the sun's diurnal motion, proceeding always from east to west, and at a constant rate, with never a slowdown along the way.

2. That of the times and places at which this daily motion starts and ends: sunrise and sunset occur at fixed times, at fixed points of the horizon, the same for the same day of the year in any two years, with seasonal variations whose pattern is invariant from year to year.

3. That of the size of the sun and of the outflow of its radiant energy: these too remain constant, so that for any given day what it gives off in light and heat will be the same from year to year.

Now let us note that regardless of the extent to which other *physiologoi* might differ from Heraclitus, every single one of them would join him in two affirmations:

1. Solar regularities are either themselves absolutely un-

breachable or else any given breach of them will admit of a
natural explanation as a special case of some other, still more
general, regularity which is itself absolutely unbreachable.
2. What makes the world a cosmos is the existence of such
highest-level, absolutely unbreachable, regularities.

Let me explain my reason for putting that first affirmation
in disjunctive form. Most of the *physiologoi*—probably all except Heraclitus himself—would have denied that the regularities of the sun's behavior are so inexorably fixed that they can
never fail. An obvious example would be an eclipse. This
could be regarded as a temporary stoppage of the radiant
energy that streams out from the sun: this is in fact Anaximander's explanation, as a blocking of the aperture in the
sun ring from which normally light and heat rush out. Heraclitus' theory is that, on the contrary, the sun is a bowl-
shaped container filled with fire, and that eclipses occur when
the bowl turns away from the earth.[12] This would enable him
to hold that the outflow of heat and light from the bowl remained constant, eclipse or no eclipse, and that the reason
why the sky was darkened in an eclipse is that while this lasted
the sun's light was emitted in another direction, away from
the earth, instead of toward it. But this sort of difference of
opinion would leave undisturbed the conviction which Heraclitus shared with Anaximander and with every other believer in the cosmos: that the failure of this, or any other,
observable regularity implies no disturbance of the more massive constancies which constitute the order of nature and, if
known, would yield the ultimate explanation of every natural
phenomenon, no matter how unusual and surprising. On this
point Heraclitus, Anaximander, and all the *physiologoi* would
stand united—a handful of intellectuals against the world.
Everyone else, Greek and barbarian alike, would take it for
granted that any regularity you care to mention could fail,
and for a reason which ruled out *a priori* a natural explanation of the failure: because it was caused by supernatural intervention.

Let us look at a couple of examples: Herodotus (7.37.2)

---

12. On Heraclitus' theory of the sun see Kirk, *Heraclitus*, pp. 269ff.

relates that as Xerxes' army was starting off from Sardis for the invasion of Greece, "the sun, leaving his place in the heavens, disappeared, though there were no clouds and the sky was absolutely clear; and there was night instead of day." It is not only Xerxes and his courtiers who take the eclipse as "an apparition" (φάσμα), a sign from the gods. So does Herodotus himself. Child of the enlightenment though he was,[13] he kept a foot in both worlds; time and again in his story of the Persian wars we see him underwriting some tale of divine intervention—"god" or *a* god causing a flood, for instance, flooding the land, or causing a savage gale (8.129; 8.13).[14] In this case, though he does not actually say so, the language he uses strongly suggests the supernatural view of the eclipse: no naturalistic view, no matter how eccentric, could have involved the sun's "leaving his place in the heavens," and doing so out of a blue sky (i.e., without winds or clouds, which might have suggested a natural explanation).[15]

The same language is used by Pindar in his Ninth Paean:

Far-seeing beam of the sun, mother of sight,
Star supreme, robbed from our sight by day,
Why have you made impotent the power of man, the path
   of wisdom
By rushing forth on a darkened path? . . .

13. For his links with *physiologoi* and sophists see, e.g., W. Nestlé, *Vom Mythos zum Logos*, the chapter on Herodotus, pp. 503ff.

14. In 8.129 the Persians attacking Potidaea are overwhelmed by an unusually high tide which floods the swamp they are trying to cross; Herodotus says he agrees with the Potidaeans that this must have been caused by Poseidon because the Persians had desecrated his temple. In 8.13 a savage gale wipes out a contingent of the Persian naval force off Euboea; "all this was done by god (ἐποιέετό τε πᾶν ὑπὸ τοῦ θεοῦ) in order that the Persian fleet might not greatly exceed the Greek, but be brought nearly to its level." For other examples of references to miracles and portents by Herodotus with no apparent skepticism see, e.g., Nestlé, *Vom Mythos zum Logos*, p. 505, no. 9.

15. How different is the language he uses when doing his own homespun *physiologia* one can see from his discussion of the summer risings of the Nile (2.24.1 and 2.53.3): he speaks of the sun in Egypt as "driven from his old [i.e., established] course by storms" during the winter and "returning to the middle of the sky" as the winter comes to an end. Here there is not the slightest hint of supernatural intervention: he is evidently thinking of the strong winds in the winter as the natural cause of the variation in the sun's customary behavior.

And here the interpretation of the event as a prodigy is explicit and long-winded:

> Are you bringing a sign of war,
> Or of failure to the crops,
> Or of an excessively violent snow-storm,
> Or of dire civil strife,
> Or of the emptying of the sea,
> Or of a freeze-up of the soil,
> Or of a heat-wave from the south
> Streaming with raging rain?
> Or will you, by flooding the land,
> Set a new start for the human race?

Convinced that the order of nature has been breached at this point, Pindar feels that almost anything can happen now: heat wave or frost are equally possible, and any number of other calamities, from crop failure and floods to war and civil strife. His conviction that it would be hopeless to look for any sort of rational explanation is voiced plainly when he asks, "Why have you made impotent the power of man, the path of wisdom?" Here "wisdom" (σοφία, intelligence) stands for the human power of rational explanation and prediction: this is what has been stymied, rendered "helpless," left without resource. All you can do in such a pass is to turn to the soothsayers—to those whose wisdom is "divine."

Now at last we are within sight of understanding why Heraclitus, and all the *physiologoi,* would have good cause to think that their discovery of the cosmos would revolutionize men's outlook. If you believe that the gods have power to make the sun drop out of his course in the sky, there will be no limit to the number of ways in which you will credit them with the ability to break into your world; and this will affect your whole attitude to what is going on around you and even, as I shall explain directly, to what is going on in you, in your very thoughts and feelings. That the area of possible divine intervention is traditionally unlimited becomes obvious when we recall the variety of things the gods were expected to do in a relatively routine matter, like policing an oath. Here is the formula of the oath of the Amphictyonic league, quoted in a speech by Aeschines (*Against Ctesiphon*, 111) in 330 B.C.:

> The curse is that their land bear no fruit;
> that their women bear children which are not like
>   their parents, but monsters;
> that their flocks produce no offspring according to
>   nature;
> that they be beaten in war, in law court, and in the
>   market;
> that they perish utterly, themselves, their household,
>   and their clan.

Apollo, Artemis, and Athena Pronaea are being credited with the ability to manipulate not only the reproductive process in man, beast, and land, but also men's mental processes. For only so could they have been expected to punish a perjurer by bringing on him a streak of losses in the law courts and in the market: to do this the gods would either have to tone up the wits of his adversaries and competitors, so that they would be more likely to come out on top in litigation and business deals, or else to impair those of the perjurer himself so as to make him do the foolish things that would prove his undoing. The latter is the course of which we hear most often in Greek literature. The phenomenon is labeled *atē*—a word whose primary sense is "bewilderment, infatuation, caused by blindness or delusion sent by the gods," with "bane, ruin, disaster" as an extended, derivative, sense. E. R. Dodds has a full account of *atē* in his beautiful book, *The Greeks and the Irrational*—an account which has already become a classic. Acknowledging the great debt I owe it myself, I must none the less register some differences from Dodds, for they affect importantly the theme of the present chapter.

First of all, belief in *atē*, though more conspicuous in the Homeric poems than anywhere else in Greek literature, is by no means as centered in the heroic and archaic age as Dodds seems to suggest. It is a permanent feature of the traditional Greek world-view; it persists into the fifth and even the fourth centuries, so that it is held by contemporaries of the *physiologoi*, and of Plato and Aristotle as well. Thus the Athe-

nian orator, Lycurgus, a contemporary of Aristotle, sub-
scribes to it himself, and expects that it would be widely
shared by his public, or he would not have risked alienating
the jury (501 Athenians selected by lot) by building it into his
speech *Against Leocrates* (92):

> The first thing the gods do to wicked men is to subvert their
> reason. I value as I would an oracle these lines, bequeathed by
> ancient poets to posterity:
>
> > When the wrath of the gods is deployed to
> >   someone's harm
> > The first thing it does is to take away good sense
> > From his wits, and give his judgment a worse turn,
> > So that he knows not when he errs.

Sophocles was no less sure of the popular acceptance of this belief in his own time; he puts it into the mouth of the fifteen The-
ban elders who form the Chorus in the *Antigone* (vv. 621–24):

> Wisely someone uttered the famous word
> That evil at times will be thought to be good
> By one whom God draws to *atē*.[16]

And now for my second, and more important point: belief
in *atē* embroils the believer in a logical predicament whose
seriousness does not seem to be fully appreciated by Dodds
and other classical scholars who have discussed this topic in
recent years. This is the crux of it: believing that someone is
acting under the influence of *atē* is logically incompatible with
holding him responsible for his action. To hold a man responsible for an act, it must be *his*—it must be the action chosen by someone with his disposition, his perceptions of the
world and of himself, his fund of memories, his present desires and aspirations. But an act described as done under the
influence of *atē* cannot satisfy this condition: the whole point
of so describing it is to call attention to the supposed fact
that, but for the intervention of the deity, the agent *would not
have acted in that way*. How then could we speak of it as his

16. Dodds, *The Greeks and the Irrational*, p. 49, quotes these lines (and the
whole of the lyric in which they occur) to illustrate "the beauty and terror of
the old beliefs." I know of no reason to believe that they would be denied
contemporary acceptance (the *Antigone* was probably produced in 441 B.C.).

act? Yet speak of it as his we must, since it is not only physically his own behavior, but psychologically as well: *atē* is not supposed to have turned him into a human puppet, going through motions without accompanying perceptions, desires, and intentions; the intruding deity is not supposed to be just moving the man's muscles and bones, but to have reached into his innermost being. His private experience— his feelings, aspirations, his sense of good and evil[17]—the whole of his psyche is supposed to have been invaded by the deity, so that the agent's mind is at that moment just what the god has wanted it to be. So the act *is* the man's own: he says, "I do thus and so," and later, "I did it," and others say, "You are doing it, you did it." Yet it is *not* his, since *he would not have chosen to do it* if he were acting in accordance with those habits and purposes which characterize him, those established personality traits, interests, and commitments which constitute his social identity. Here then is a muddle, a contradiction, which cannot be avoided, or even palliated, in either of the two ways which have proved popular among scholars:

One of these is the line taken by Dodds, followed with modifications by A. W. H. Adkins in his influential books.[18] This line is predicated on the claim that archaic morality and jurisprudence impute responsibility solely on the basis of the act with no regard for what lawyers call *mens rea,* that is, the mental state of the agent. "Early Greek justice," writes Dodds, "cared nothing for intent; it was the act that mattered."[19] If this claim could be made good the contradiction would be resolved, for then the question of whether an act was chosen by the agent or forced on him by a god or goddess could be ignored: on that hypothesis he could be held responsible for the act simply because he did it. I submit that this resolution will not work, for the claim on which it is

17. Cf. the citations from Aeschines and Sophocles in the preceding paragraphs of the text above.
18. A. W. H. Adkins, *Merit and Responsibility; From the Many to the One; Moral Values and Political Behaviour in Ancient Greece.*
19. Dodds, *The Greeks and the Irrational,* p. 3.

based is demonstrably false. There are cases in both the *Iliad* and the *Odyssey* in which a man is cleared of responsibility when it is established that the offending act was done in ignorance or under duress. For example, in Book 22 of the *Odyssey*, when Odysseus has just slaughtered the suitors and is about to do the same to the bard, Phemius, for having connived in their *hybris*, Phemius pleads that he had associated with the suitors only under duress, hence "with no will or desire of [his] own" (οὔ τι ἑκὼν . . . οὐδὲ χατίζων, v. 351). He calls on Telemachus to corroborate his plea; and when Telemachus does so, Phemius is absolved, declared free of culpable responsibility (ἀναίτιος, v. 356); he goes scot free.[20] By the same token everyone who believes himself, or is believed by others, to have acted under the influence of *atē* would have to be cleared of responsibility, since *atē* would be even stronger—totally irresistible—duress. This is in fact how Agamemnon in Book 19 of the *Iliad* represents that act of his which had wronged Achilles and had provoked the hero's "baneful wrath":

> I am not responsible (ἐγὼ δ'οὐκ αἴτιός εἰμι)
> But Zeus is, and Fate, and the Erinys that walks in darkness:
> These put into my wits a savage *atē* in the assembly
> When I despoiled Achilles of his prize.
> So what could I do? Deity will always have its way
>   [Vv. 86–89].[21]

But nonetheless he is held responsible and he holds himself responsible and proceeds to make amends for his misdeed exactly as if he had been, and had thought himself, responsible. This is the contradiction I have been talking about, and you can see it here stark and unresolved.

Nor can this contradiction be resolved by way of the principle, now widely accepted among classicists, that in the case of *atē* a double causation—by the god *and* by the human agent—is being supposed, hence dual responsibility.[22] This

20. For another example see Appendix, section A.
21. The translation owes much to that in Dodds, *The Greeks and the Irrational*.
22. A. Lesky, "Göttliche und menschliche Motivation im homerischen Epos," followed, e.g., by H. Lloyd-Jones, *The Justice of Zeus*, pp. 9ff.

principle is indeed true,[23] and we see it dramatically instanti-
ated in the Embassy scene in the *Iliad* (Book 9): there, in a
speech by Ajax, which is in every other way sane and wise,
Achilles' wrath is represented, first as what *he*, Achilles, "has
put into his breast," and then, just eight lines later, as what
"the gods put in Achilles' breast." The contradiction is bla-
tant. But neither the speaker, nor Achilles, nor anyone else
there makes any allusion to it: it is just one aspect of the
predicament that all who believe in *atē* have learned to live
with. If you are struck by *atē*, the god who strikes you puts
thoughts and feelings into your heart—yet you too put them
there, for otherwise the resulting act would not be yours, and
what the god did to you would not be *atē*.

I have dwelt so long on *atē* because it dramatizes the perception of an irrational universe which the *physiologoi* inherited as did everyone else in their time. But, of course, it is
only one of the many possible examples. For just one more,
I shall revert to the perception of an eclipse, illustrating by
the most famous and fateful eclipse in the whole of Greek
history. The scene is the plain before Syracuse, where the
Athenian expeditionary force and that of its allies—the largest
ever committed by Athens in a military engagement—has
been encamped. Its position has become indefensible and immediate withdrawal is mandatory: further delay would be
criminal folly. Let Thucydides continue the story:

> The preparations for the departure were made and they were on
> the point of sailing, when the moon, being just then at the full,
> was eclipsed. The mass of the army was greatly moved and called
> upon the generals to remain. Nicias himself, who was too much
> under the influence of divination and the like, refused even to
> discuss the question of the evacuation until they had remained
> 27 days, as the soothsayers prescribed. This was the reason why
> the departure of the Athenians was finally delayed. [*History*, 7.50.4]

And the result? The whole of Nicias' army was wiped
out—killed or sold into slavery. Nicias was not an ignorant

---

23. I.e., as a description of Greek belief. What I am denying is only that it
helps in the slightest to resolve, or even mitigate, the contradiction in the
state of mind which it reports.

man. He must have fancied himself an intellectual. We see
him in one of Plato's dialogues, the *Laches*, taking a leading
part in a Socratic elenchus, acting as a partisan of the highly
intellectualized Socratic view of courage (194Dff.). But he was
still a prisoner of the traditional world-view where the sun
and the moon *could* "overstep their measures" and would at
the behest of supernatural powers.

In using the term "supernatural" here, and a little earlier,
to identify the factor whose agency made the crucial difference between the world of traditional belief and the cosmos
of the *physiologoi* I may be exposing myself to the charge of
anachronism: Nicias and his soothsayers could not have said
that the eclipse had a supernatural cause—they had no such
word. That compound has no counterpart in classical Greek:
hyperphysikos is not in Nicias' or Thucydides' or Plato's language. But what their language does have are the counterparts—indeed the true originals—of one of that term's components: the words for *natural*, and *nature, physikos, physis*.[24]
And this gives me all I need. For *physis* is the key term in the
transition from the world of Homer, Hesiod, Archilochus,
Sappho, Aeschylus, Sophocles, Aristophanes, Herodotus,
and the orators—that is to say, from the world of common
belief and imagination throughout the archaic and classical periods—to the world of the *physiologoi* and of a few
tough, hard-bitten, intellectuals like Thucydides and the Hip-
pocratics—the world which was cosmos. That *physis* is even
more basic than *kosmos* is evident from the fact that the discoverers of the cosmos came to be called *physiologoi*, not *kos-
mologoi*, and that "nature" occurs much more frequently in
titles of their treatises than does "cosmos."[25] The beauty
of *physis* is that it shows what it was in the traditional conception of the world that gave the *physiologoi* the building materials for their new construction. The cosmos they had to
invent. *Physis* they found ready-made in the inherited con-

24. For a fundamental study of the concept of *physis* in early Greek
thought see W. A. Heidel, "Peri physeōs." For an important subsequent
monograph which collects illustrative passages see D. Holwerda, *Physis*.
25. For the titles see the Word Index in Diels-Kranz, Vol. 3, *s.v. Physis,
Kosmos*.

ceptual scheme; all they needed to do with *physis* was to make a new use of it.

To explain this, let me point out how *physis* is used by Herodotus. In his prose the *physis* of any given thing is that cluster of stable characteristics by which we can recognize that thing and can anticipate the limits within which it can act upon other things or be acted upon by them. Thus in his chapter on Egypt, Herodotus introduces his Greek readers to two of its unfamiliar animals through descriptions which begin, "The *physis* of the crocodile is of this sort . . ." (2.68.1); "The hippopotamus has a *physis* of this form . . ." (2.71). Now from the fact that a given thing has a *physis* Herodotus would not allow us to infer that we will always see it in full possession of that *physis*. Thus it is the *physis* of a crocodile to have a tail; but it does not follow that this crocodile will have one; he may have lost it in a fight, or in some other way. The one thing that is certain for Herodotus is that, barring supernatural intervention, whenever things interact their *physeis* set limits to what can happen. Thus when he hears the story that Hercules, when he came to Egypt, was seized by a mob and marched up to an altar to be sacrificed, whereupon "he put forth his strength and slew them all" (2.45.3), Herodotus asks, "If Hercules was a man, as they admit, how would it be natural (κῶς φύσιν ἔχει) for [this one man] to kill many thousands?" Note the force of the conditional. If Hercules were not a man, but the demigod he became after his death, that question would not have been raised. In this Herodotus stays within the traditional view, where *physis* fixes the limits of the possible for everything *except* the supernatural.

What the *physiologoi* do is to drop that exception. They make the world a cosmos by keeping what was already there in the form of *physis* and cleaning out everything else. They do so without saying so. They could not say so in a society whose public cult, saturated with the supernatural, was ensconced in the state establishment and enjoyed the protection of the law. To attack the supernatural head-on would turn them into outlaws. So they do the next best thing. They proceed by indirection. They so fill up the universe with *physis* as to leave no room for anything else. This is a two-pronged operation. The first prong invades the heartland of the supernatural: the heavens, whose sun and moon had been thought gods; and the *meteōra*, between the heavens and the earth, whose rain, clouds, lightning and thunder, mists, winds, rainbows, storms were so closely associated with divine control that the formulaic epithet for Zeus in Homer had been "cloud-gatherer" and the Greek expansion of "raining" would be, not "it is raining," but "he—Zeus—is raining." The other prong of the operation produces theories of the *physis* of the universe as a whole. These are at first genetic (they tell of a primordial substance from which our world and any others, if such there be, arise, and to which they eventually return), but then become increasingly theories of the invariants of change (fire, water, earth in Heraclitus; the dark and bright forms in the cosmology of Parmenides; the four "roots" in Empedocles; the infinitely numerous qualitatively multiform seeds in Anaxagoras; and the infinitely numerous qualitatively uniform atoms in Leucippus and Democritus).

On both of these operations the demolition of the supernatural is accomplished without a single word about the victim. In the first the job is done just by implementing a single tacit axiom of rational explanation: that whatever happens in any region of the universe—in the heavens, beyond it, or under it—involves only interactions between material entities whose *physis* is always the same wherever it turns up and conforms to the same highest-level regularities. So long as that axiom is observed theories of the same phenomena may vary wildly while making the same contribution to the perception of a rational universe. What could be more different than the two theories of the sun we noticed above: Anaximander's, that the sun is a huge annular body filled with fire, revolving around the earth throughout the years, all of it totally invisible except for a small part of it, an orifice from which fire keeps streaming out—and Heraclitus', that the sun is a bowl-shaped container, containing fire "maintained by moist exhalations or evaporations from the sea, which are somehow collected in [it] and burned as fuel."[26]

And there were other theories too, as different from the one as from the other—Anaximenes' sun, for example, a disk of fire, "flat as a leaf," floating on air, as durable as Anaximander's, but no larger than was the aperture in the latter's fire-ring. Anaximander's sun impresses us by its imaginative grandeur; the suns of Anaximenes and Heraclitus depress us by their crudeness and naïveté. But the fact is that while in none of the three cases do we have anything we can begin to call a *scientific* explanation of the sun, all three provide an explanation that would be *rational* in a comfortably pragmatic sense (any one of the three would have sufficed to put out of business soothsayer specialists on eclipses) and also in a more reflective sense (highest-level natural uniformities would be secure in all three; the fire in each would be as certain to behave according to its *physis* as would the fire in your hearth).

In the second operation the spectrum of variations is even wider. Although I have mentioned most of the major ones, I cannot bring this chapter to a close without alluding to two more which are particularly significant for Heraclitus' place within that spectrum. On one hand, there is the question of whether or not the orderliness of the physical universe should be described in moral terms. In the earlier phase of *physiologia* we see strong answers in the affirmative. So, for example, in Anaximander (fragment B1), who sees in cycles of natural changes "injustice" followed invariably by reparation.[27] So too in Heraclitus to whose faith that "strife is justice" I referred above. But two generations later, in the atomists, the universe is as severely nonmoral as the human imagination has ever made it. Leucippus and Democritus would no more think of reading justice into the cosmos than would

---

26. Kirk in G. S. Kirk and J. E. Raven, *The Presocratic Philosophers*, p. 203.

27. B1. For my interpretation of this much discussed fragment see my "Equality and Justice in Early Greek Cosmologies" in Furley and Allen, eds., *Studies in Presocratic Philosophy*, 1:73ff.

Thucydides or Camus. On the other hand, there is the question of whether or not the *physiologoi* can find a home for deity within their cosmos. Some do and some do not. Heraclitus apparently does. He writes: "The One, the only wise, will and will not be named 'Zeus'" (B32). It will not, because it is the fire dispersed throughout the universe in measured kindlings and extinguishings. Yet it will, for this same fire is somehow also mind—"the thought which steers all things through all things" (B41)—and thus discharges in the universe, though on a vastly grander scale, the role of supreme ruler traditionally ascribed to Zeus. And there are others, like Anaxagoras and Diogenes of Apollonia, who postulate a mind that is ultimately responsible for the order which makes a cosmos of the world; the latter speaks quite explicitly of this cosmic intelligence as "god." At the other extreme, we have the rigorous materialism of the atomists, where order is inherent in matter, and an ordering cosmic mind is too incongruous a notion to be even worth arguing against.

What unites these men, in spite of these and other differences, is that for all of them nature remains the inviolate all-inclusive principle of explanation. In the Heraclitean cosmos a god-sent eclipse in the heavens is as unthinkable as in the Democritean. *Atē* is equally unthinkable in both. Heraclitus can say, "a man's fate is his character" (B119), with the same conviction as would Democritus. For the first time in history man has achieved a perception of a rational universe which leaves his own destiny to be determined solely by *physis*—his own and that of the world.

2
Plato's Cosmos, I:
Theory of Celestial Motions

IN Book 10 of the *Laws* Plato drafts a statute against impiety which is without parallel in any surviving code of ancient Greece.[1] The mildest of its penalties is five years' solitary confinement, to be followed by execution if the prisoner is still unreformed. A man of irreproachable character would get this sentence if it were proved that he believed that there are no gods or that they do not care for men (888C). As one reads the philosophical preamble (885Bff.) it dawns on one that the Ionian discoverers of the cosmos, the *physiologoi,* are in Plato's eyes the main fomenters of the heresies from which his Utopia must be purged. His punitive measures are directed first and foremost against men who hold that the original constituents of the universe are material entities like earth, air, water, and fire, which exist "by nature" and "by chance"—that is to say, things whose existence did not come about by design or, as Plato puts it, "by art": things which just happened to exist. And they teach that this cosmos of ours, with all of the "art" in it, is the chance outcome of the natural interactions of these soulless bodies.[2]

1. See the comment by G. R. Morrow, *Plato's Cretan City,* pp. 488–89: "Plato becomes the first political thinker to propose that errors of opinion be made crimes punishable by law . . . with fateful consequences to Western history, for henceforth the punishment of errors of opinion could claim the sanction of one of the highest authorities."

2. The crucial passage, retranslated:

By nature and by chance, they say, fire and water and earth and air all exist—none

Some of our *physiologoi*, arraigned for impiety before the Court for Capital Offences in Plato's Utopia, would have protested this account of their system: Heraclitus and Diogenes of Apollonia, for example, for, as I have pointed out, both of them endowed their universal substance with intelligence and even thought of it as a god. In other ways too we might fault the exact applicability of Plato's indictment to each of the systems of the *physiologoi*.[3] But if we see in Plato's diatribe against them not historical reportage, but a philosopher's diagnosis of the basic difference between his own world-view and theirs, what he says is profoundly true: it gets at the root of the matter. Regardless of many disagreements among themselves, the *physiologoi* are united in the assumption that the order which makes our world a cosmos is natural, that is to say, that it is immanent in nature; all of them would account for this order by the natures of the components of the universe without appeal to anything else, hence without appeal to a transcendent ordering intelligence. Plato is dead right on this point—obviously so in the case of the last and most mature product of Ionian *physiologia*, the atomistic system, where our own cosmos and infinitely many others are

---

of them exist by art—and the bodies which come next—the earth, the sun, the moon, and the stars—were generated by these totally soulless means: things moved by chance, each by its own power, as they happened to combine fittingly somehow with one another—the hot with the cold or the dry with the moist or the hard with the soft, and all those things which through the mingling of opposites chanced into forced intermixture. In such a way and by such means the whole of the heavens and everything pertaining to it has been generated, and moreover all animals and all plants, all of the seasons having been generated from these same things, not by intelligence, they say, nor by a god, nor by art, but by what we are talking about: by nature and by chance. [889B1–C6]

The awkwardness of this long sprawling period (symptomatic of stylistic deterioration in Plato's old age) leaves its sense perfectly clear. There is no problem for the translator except one: in the clause, διὰ τούτων γεγονέναι παντελῶς ὄντων ἀψύχων (889B4–5), should the last three words be taken with τούτων in the same clause (which refers to "nature and chance"), or else with τὰ μετὰ ταῦτα αὖ σώματα, γῆς τε etc. in the preceding clause? I take the answer to be: with both. Fire, water, earth, air are "totally soulless," and so are "nature and chance" (i.e., nature as conceived by rank materialists: devoid of purpose). If by διὰ τούτων γεγονέναι Plato had meant only "they were generated from those primary elements" (so Diès), the preposition would have been ἐκ, not διά.

3. Cf. my "Equality and Justice in Early Greek Cosmologies," pp. 176–77.

produced within the infinite universe by purely mechanical causes, when atoms colliding at random in the void happen to fall into a particular mechanical pattern, a vortex. Even those other systems where one of the constituents of the universe is made the producer of order and is endowed with the attributes of reason—as is fire in Heraclitus, air in Anaximenes and Diogenes—even they would bear out the Platonic diagnosis: for they too assume that order is inherent in nature and does not need to be imposed upon it by a supernatural ordering mind.[4] The difference on this point Plato considers so irreconcilable with his own view and so momentous in its moral and political implications as to justify the removal from the body politic of anyone who holds it. What does he offer in return? A theological cosmogony.

What Heraclitus had denied when he wrote, "this world, the same for all, no man or god has made," Plato makes the first principle of cosmology in the *Timaeus*. He undertakes to depict the origin of the cosmos as the work of a god who takes over matter in a chaotic state and moulds it in the likeness of an ideal model, the Platonic Idea of Living Creature (30Cff.). That this god is supernatural in the literal sense of the term is plain enough: he stands outside of nature and above it; he is not himself a member of the system of interacting entities which constitutes nature; he acts upon that system, but the system does not act on him. But how far would Plato go in assimilating this god to the supernaturals of popular belief? Would this divine world-creator have the power to violate the regularities of nature? The answer is not clear. What *is* clear is that, if he does have such power, he will

---

4. For Heraclitus and Diogenes see above p. 22. The real exception, of course, is Anaxagoras. He starts off the cosmogonic vortex with an extramundane "Mind" which exists apart (B12) from matter in its primordial, undifferentiated, state, yet somehow manages to get the vortical motion started (B13) and even to "dominate" it (B12) thereafter. Plato knows all about this effort by a *physiologos* to make Mind the ordering agency of nature. But he indicts the effort as abortive (*Phaedo* 97C–99C): in spite of that attempt, Anaxagoras' cosmogony remains for all practical purposes mechanistic, explaining the main features of the universe by material causes, without reference to the purposes Mind was bent on realizing in and through these mindless things.

never choose to exercise it. This follows, I think, from the conception of his character in the *Timaeus*.

It would be hard to think of two more sharply divergent ideas of the supreme deity than Plato's and the traditionalists'—this in spite of the fact that intelligence is a dominant attribute in each case. Zeus had always been thought brainier than the other gods, with the possible exception of Athena who, after all, had been his brain-child. But the Olympian's wisdom had been that of a deep-scheming, far-sighted, monarch, while the high god of the *Timaeus* is not so much a governor as a philosopher, a mathematician, an engineer, and, above all, an artist. That he is Reason personified is taken for granted; Plato alludes to this from time to time, but feels no need to say so formally. He does not call his deity "Nous" or "Logos," but "Demiourgos"—literally, "Craftsman." The name is surprising. In Plato's Athens the craftsman is often a slave and as often a freeman working shoulder to shoulder with slaves in the same kind of work. So the scorn which the leisure classes feel for the slave tends to rub off on the craftsman too: he becomes the victim of stigma by association, and in all but the most advanced forms of democracy his civic status is precarious. Plato denies him political participation in the *Republic* and robs him even of citizenship in the *Laws*. So does Aristotle in his "polity" which is meant to be a very moderate constitution, a "mixed" one, a compromise between aristocracy and democracy. That the supreme god of Plato's cosmos should wear the mask of a manual worker is a triumph of the philosophical imagination over ingrained social prejudice.

To be sure the images of "king" and "father" are also invoked; but they are marginal—traditional hang-overs that do not dictate the working imagery of the dialogue. Neither the sexual nor the domestic associations of paternity are followed out in the creation story, nor yet the political ones of kingship, but only those prompted by the Craftsman metaphor. It is phrases like "contriving," "moulding," "measuring out," "pouring into a mixing-bowl," "cutting up," and "splicing together" that carry the bulk of the narrative. But this divine mechanic is not a drudge. He is an artist or, more precisely, what an artist would have to be in Plato's conception of art: not the inventor of new form, but the imposer of pre-existing form on as yet formless material. The Demiurge chooses the Idea of Living Creature for this purpose because he has the artist's longing to create a thing of beauty, and that model is the most beautiful he can find; and, further, because, says Plato, "he was good and in the good there can be no envy at any time about anything" (29E).

Here is another striking deviation from established patterns. The envy of the gods had been one of the deepest convictions in Greek theology, and one of the most persistent: it is more evidence in Herodotus and Euripides than in Homer.[5] Greek gods are so defensive of their privileged status *vis à vis* men that they grudge humanity any benefit, however innocent in itself, that would narrow the gap between the splendor of their own existence and the wretchedness of the human estate. To exalt themselves they abase men. That is why they punish Prometheus so savagely. His motive, *philanthropia*,[6] does not extenuate his crime in bringing man the gift of fire, but rather aggravates it. The Olympians can no more forgive this "man-loving" god than an aristocratic white caste could forgive one of its members who turns into a "nigger-lover." That is why Plutus, the god of wealth, is blinded by Zeus in Aristophanes' comedy.[7] While

---

5. For Herodotus and Euripides see, e.g., S. Ranulf, *The Jealousy of the Gods and Criminal Law at Athens*, chap. 5. For brief and penetrating comment see Dodds, *The Greeks and the Irrational*, pp. 30–31.

6. That this is what he is being punished for is stressed from the start of the *Prometheus Vinctus* of Aeschylus: his theft of the fire he brought to mortals is "the sin for which he must pay just penalty to the gods that he may learn to welcome the tyranny of Zeus and cease his man-loving disposition" (vv. 9–11).

7. *Plutus*, vv. 86ff. Dodds points out that in writers like Aeschylus and Herodotus "divine *phthonos* [envy] is sometimes though not always, moralized as nemesis, 'righteous indignation'" (*The Greeks and the Irrational*, p. 31 and notes). Lloyd-Jones goes further; he holds that for these and other fifth-century writers divine *phthonos* "actually formed part of justice" (*The Justice of Zeus*, pp. 69–70). I cannot follow him on this point. That the enlightened spirits of the age should do their best to sublimate a primitive concept is understandable. But the material in this case was intractable. The *Plutus*

he could see, he brought prosperity only to good and wise men. Zeus blinded me, says Plutus, because he was "envious of mankind": Zeus could not tolerate the improvement of the human lot that would ensue if Plutus were allowed the ability to insure that wealth will always match virtue in man's life.

Unlike the cruel tyrant of the *Prometheus Vinctus* and the *Plutus*, Plato's Craftsman is driven by the desire to share his excellence with others; the more beauty and goodness outside of him, the better his unenvious nature is pleased. This is the noblest image of the deity ever projected in classical antiquity and it opens the way to a radically new idea of piety for the intellectual which the traditionalists would have thought impious: that of striving for similitude to God (ὁμοίωσις θεῷ). If I were in a position here to trace out the implications of this idea, I think I could show how inspiring it is and yet disquieting, for it connects with the ominous notion of the philosopher king in the *Republic*. But here my concern is not with Plato's contribution to religion and morality, but to cosmology. So let me proceed to that.

The accomplishments of the Demiurge fall into two categories:

I. Triumphs of Pure Teleology (Part I of the Creation Story, 29E–47D).

II. Compromises of Teleology with Necessity (Part II, from 47E to the end).

Category I comprises two classes of propositions: those which admit of teleological *derivation,* on one hand, those

---

passage (which is ignored in Lloyd-Jones, as are also the lines I quoted from the *Prometheus Vinctus* in the preceding note) shows it surviving unreformed: so far from being "a part" of justice, here the *phthonos* of the gods operates to defeat the just apportionment of human wealth. When Aeschylus contrasts his own conviction (emphasizing how personal and unrepresentative it is: "apart from others, alone in my thought") that it is not prosperity as such, but wickedly won prosperity, that is struck down by the gods (*Agamemnon,* vv. 757–62, with the comments in J. D. Denniston and Denys Page), he does not employ the idea of divine envy in a purified, reconditioned, form. He ignores it, and with good reason: the very notion of just gods *envying* unjust aggrandizement is irredeemably perverse. For *phthonos* as a passion whose "very name entails badness," so that it is always bad, without acceptable "mean," see Aristotle, *Nicomachean Ethics* 1107A9ff.

which admit of teleological *explanation*, on the other. To begin with the first, here Plato startles us, claiming to discover facts about our universe by deducing their existence from the postulates of his cosmology, that is, from the hypothesis that the universe is produced by a god whose benevolence and love of beauty prompt him to create a physical likeness of the Ideal Form of Living Creature. Here are some of these supposed facts:

1. The cosmos must have a soul. Why so? Because the Ideal Form of Living Creature has a soul (30B).

2. The cosmos must be unique. Why so? Because the Ideal model is unique, and the world would be more like that model if it too were unique (31A–B).[8]

3. The cosmos must be spherical. Why so? Because the sphere is the most homogeneous (ὁμοιότατον) shape there is, and the homogeneous, says Plato, is "ten thousand times more beautiful" (33B7) than the heterogeneous.

4. The cosmos must be characterized by time. Why so? Because time, as Plato thinks of it, is "a moving image of eternity" (37D5); it brings a dimension of ordered constancy to the inconstancy of flux, and thus makes the changing image more like the unchanging model than it would otherwise be.

I am not disposed to linger on this part of the creation story. But a historian would be derelict in his duty if he failed to acknowledge the retrograde turn which Plato gives to cosmological inquiry when he converts so blatantly preconceptions of value into allegations of fact. To be sure, he was not the first to do this. Pythagoras had done so openly long before.[9] Even the *physiologoi* were not completely immune to that tendency: Anaximander and Heraclitus, as we saw in the preceding chapter, had projected into the physical cosmos their faith in justice. But there is a vast difference between their practice and Plato's. Suppose we could have asked the *physio-*

---

8. For a curious error in Plato's reasoning see D. Keyt, "The Mad Craftsman of the *Timaeus*," pp. 230–35.

9. On this feature of his system all scholars would agree in spite of their disagreement over his other doctrines.

logoi—any of them, from Anaximander down—the following
question: 'In your inquiries into nature, when you have to
decide whether or not something is thus and so, would you
think it right to settle the issue by arguing, "It would be better, more beautiful, if things were thus and so; *ergo*, they are
thus and so?"'[10] There can be little doubt, I think, that we
would have got an emphatic, 'No,' from each. Certainly that
is how Plato thinks of them: he denounces them in the *Phaedo*
(97Cff.) for failing to invoke the principle of the good in their
natural inquiries; he reproaches them for deciding such a
question as whether the earth is flat or round without first
asking which of the two would be the "better" (97E). It would
have gone against the grain of Ionian *physiologia* to concede
such a principle; and in its final, atomistic, chapter the principle is ejected root and branch. When Leucippus declares,
"Nothing happens at random, but everything from reason
and necessity" (B2), the "reason" to which he refers is the
reason *of* necessity—the reasoning of mechanistic explanation
which excludes rigorously and systematically considerations
of value from the determination of matter of fact. This is, in
good part, why the atomistic system proved so congenial to
the founders of modern science in the seventeenth century—
why they welcomed it as a haven from the Aristotelian system
which had consolidated the teleological methodology champi-
oned by Plato.[11]

But there is more to this part of the *Timaeus*. Plato fortu-
nately does not stay stuck in this shady area where one gets
one's facts by deducing them from theological and metaphysical premises. He soon comes within sight of *facts derived from
a scientific discipline*, and then, with those facts in hand, in-
vokes the teleological framework of his creation story to
structure them in a coherent scheme. Let us look at this
scheme. In some ways it is more fantastic than anything we
have seen so far. But let us not prejudice it until we have seen
what is the work which it will do for Plato.

10. I put first-instance imaginary quotations in single quotes, reserving
first-instance double quotes for quotations from cited sources.
11. Cf. below p. 62 and n. 102.

Since a soul must be provided for the cosmos, the Demiurge proceeds to its creation. Plato now implements one of his long-standing tenets, namely, that soul straddles the two disparate realms of his ontology—that of the abstract Forms, on one hand, which constitute the world of eternal Being, and that of sensible things, on the other, which constitute the ever-changing world of Becoming. Soul has a leg in each of these. In rational thought, whose objects are pure Forms, it has contact with the world of Being. In sense-perception, whose objects are changing things, it has contact with the world of Becoming. Plato, therefore, pictures the creation of soul as a blending of Being and Becoming.[12] Mixing these up in his mixing-bowl, the Demiurge produces a new kind of stuff—or, should I say, super-stuff?—which contains no physical matter (no fire, air, water, earth) and has none of the properties of physical matter (such as temperature, density, weight) except one: it can move. But even in its capacity for motion it differs radically from physical matter. For Plato the latter is inert—any part of it moves only when it is moved by something other than itself; soul, and only soul, can move itself:[13] by thinking and willing it can move the body to which it is attached and, through this, other bodies. So in creating soul the Demiurge does something which will have vast physical consequences: the self-caused movement of the World Soul and of the souls of the stars will account for every movement in the heavens: all celestial motion is to be explained as psychokinesis.

The Demiurge creates the World Soul by cutting up into strips the soul-stuff he has produced, and joining the ends to make "circles," that is, mobile circular bands. Why circular? Because of another Platonic doctrine: that rotary motion is the one "most appropriate for reason and intelligence" (34A); only thus, Plato thinks, can the absolute invariance of the eternal Forms be approximated within the ceaseless variance which is inherent in motion. The first of the soul-circles the

---

12. I take this to be the main point of the complicated psychogony at 35Aff.
13. Self-motion is said to be "the essence and definition" of soul in *Phaedrus* 245C–E; soul is "the thing that is self-moved" in *Timaeus* 37B5.

Demiurge puts at the circumference of the cosmos to produce "the movement of the Same," a movement which Plato
thinks runs through the whole universe: everything in the cosmos, from its extreme periphery down to the center of our
earth, is subject to this motion,[14] though this is counteracted
by other motions everywhere[15] except in the region of the
fixed stars. In their case we can see it pure and unimpeded:
it is the diurnal revolution from east to west parallel to the
plane of the celestial equator[16]—the World Soul's self-caused
"movement of the Same," communicated to each of the fixed
stars, so that it becomes *their* motion in turn.[17]

But Plato must also account for another set of celestial motions: those peculiar to the "wandering" stars,[18] the moon,
the sun, and the five planets.[19] Each of these exhibits a long-

14. He speaks of the movement of the Same as "everywhere inwoven from
the center to the outermost heaven and enveloping the heaven all around
on the outside" (36E2–3, Cornford's translation).

15. It follows that it has to be counteracted in the earth, for otherwise the
earth would share the diurnal motion of the fixed stars, and this would can-
cel out their motion relative to the earth. Just how this counteracting is accomplished in this case is a matter of long-standing controversy which I need
not enter. The reader may consult such treatments of the topic as those in
T. L. Heath (*Aristarchus of Samos, the Ancient Copernicus*, pp. 174ff.); F. M. Cornford (*Plato's Cosmology*, pp. 120ff.); H. Cherniss (*Aristotle's Criticism of Plato and
the Academy*, Appendix 8, pp. 540ff.); D. R. Dicks (*Early Greek Astronomy to
Aristotle*, pp. 132ff.); W. Burkert (*Lore and Science in Ancient Pythagoreanism*, p.
326, n. 16). For references to the older literature see Heath and Cherniss.

16. Speaking, as always in this chapter, from the geocentric viewpoint of
ancient astronomy.

17. This is their "forward" motion; as each is "mastered by the revolution
of the Same and uniform" (40B). They have an additional motion which is
all their own: axial rotation ("uniform motion in the same place," 40A).

18. For the import of Plato's use of this expression in the *Timaeus*, see
Appendix, section B.

19. A terminological point: my use of the term "planet" will follow Plato's.
In the only two passages in which he refers to the term πλανητόν as a name
for celestial bodies—38C5–6, ἐπίκλην ἔχοντα "πλανητά," and *Laws* 821B9,
ἐπονομάζοντες "πλανητά" αὐτά—he is clearly restricting it to the five planets
of Greek astronomy (this is explicit in the *Timaeus* ["sun, moon, and five
other stars surnamed 'wanderers'"]; it is also clear in the text of the *Laws* and,
though disguised in loose translations of it, becomes clear enough when the
passage is translated strictly, as, e.g., by Diès or Apelt). Postclassical usage,
often echoed by modern scholars in discussing Greek astronomical theories,
sanctions the extension of the term to include the moon and the sun as well

term motion of its own from which the fixed stars are totally exempt. For if the successive positions of any member of the septet are plotted against those of the fixed stars over extended periods of time, they will be found to keep shifting eastward, so that eventually they will have made a complete circuit of the sky. These slow eastward orbits differ conspicuously from the diurnal westward revolution which predominates in the sky. Three of the differences are specially noteworthy:

In the first place, all of their periods are much longer, with wide variations in length among them. The moon takes a lunar month to return to her original position; the sun a solar year and Venus and Mercury the same (on the average); Mars 1 year plus 322 days; Jupiter 11 years plus 315 days; Saturn 29 years plus 166 days.

In the second place, these orbits proceed in different planes from those of the fixed stars. Instead of running parallel to the plane of the celestial equator, all of the eastward orbits of the septet move in planes which intersect the latter obliquely. Thus the sun moves in the plane of the ecliptic which intersects the plane of the celestial equator at an angle of 23.5°.[20] The moon and the five planets move in varying degrees of proximity to the ecliptic, their orbits falling almost entirely within "the Zodiac circle"—a narrow band of fixed stars on either side of the ecliptic, conveniently figured in a sequence of twelve distinctive constellations.

─────────────

(though even then "sun, moon, and the five planets" remains the usual form of designation for the septet of "wandering" stars: see, e.g., the citation from Ptolemy in n. 107 below). Since Plato himself refers so pointedly in the *Timaeus* to all seven of these bodies as "subject to turnings and wanderings" (τρεπόμενα καὶ πλάνην ἔχοντα, 40B6)—though only on the understanding noted in section D of the Appendix—he would no doubt tolerate this extended usage. But since he does not himself employ it, I prefer to stick to the narrower use of the term which is both more convenient and, after all, Plato's own.

20. Reckoned at 24° (the angle subtended at the center of a circle by the side of a regular fifteen-sided polygon) in the fourth century B.C. (so, e.g., Eudemus [fr. 145 Wehrli] *ap.* Theo Smyrnaeus, *Expositio rerum mathematicarum*, E. Hiller, ed. [hereafter to be cited by author's name only], p. 198). Plato does not refer to the ecliptic as such. But he certainly presupposes it: see n. 23 below. Its discovery antedates his birth: see nn. 42 and 43 below.

In the third place, all of these orbits exhibit the phenomenon which is so marked in the seasonal behavior of the sun: all of them have "turnings" (τροπαί), points of maximum deviation north and south at which they turn back and proceed in the reverse direction until the opposite point of "turning" has been reached.[21]

To account for all these features of "wandering" motion Plato postulates another "circle" in the World Soul which he calls "the movement of the Different"[22] and describes it as "slanting" (πλαγίαν 39A) in an inverse direction, "toward the left by the way of the diagonal," while the motion of the Same is "toward the right by way of the side."[23] This second circle is broken up into seven circles of unequal length with appropriately different angular velocities which are distributed to the moon, the sun, and the five planets to become

21. At least as far back as Hesiod (*Works and Days*, vv. 479, 564, 663), probably even in Homer (*Odyssey*, bk. 15, v. 404), "turnings of the sun" (τροπαὶ ἠελίου) is the common label for the summer and winter solstices. When the planets are discovered, the term is naturally extended to them, since they too exhibit similar "turnings."

22. Plato's term, θάτερον, could also be translated "the Other," and is often so rendered in the literature. I prefer "Different" because it is more faithful to the idea which motivates Plato's caption, that this movement represents difference, not just in the bare sense of logical nonidentity but in the more concrete sense of variety and diversification; the movement devolves into no less than seven different patterns of motion, all of them slanting eastward, but in different tempos and figures.

23. "By way of the diagonal" (κατὰ διάμετρον vs. κατὰ πλευράν, 36C). Plato has in view a geometrical diagram: a rectangle inscribed in a circle, whose upper side represents the summer tropic, its base the winter tropic, and the diagonal the ecliptic. The east is to its left, the west to its right, probably because the diagram represents the view of an observer in a northern latitude looking toward the south (for as the sequel [39D2-7] suggests, the view is toward the area of the heavens traversed by the circuits of the planets); if it were viewed from the south, the east would be on the right (so spoken of in *Laws* 760D), the west on the left. That the diagonal motion "toward the left" is so much slower than the lateral motion "toward the right" is left unmentioned, which suggests that Plato now takes it as so very obvious that it should go without saying. He felt a little differently on this point—it may have been less of an old story to him then—when sketching the world-model in the Myth of Er in the *Republic*: there he spoke of "the seven circuits [those of Sun, Moon, and five planets] revolving *slowly* (ἠρέμα) in a direction contrary to that of the whole" (617A5).

their individual, characteristic, motions, and account for the eastward movement of each in, or close to, the plane of the ecliptic.

Thus before coming around to the creation of humanity the Demiurge has brought into the world a huge population of "everlasting gods" (37B6): the World Soul; the countless multitude of fixed stars,[24] whose visible motion[25] is exclusively the movement of the Same; and the sun, the moon, and the five planets, whose visible motions exhibit also diversifications of the movement of the Different. Because of the invariant periodicity of their motions the stars provide visible measures of time: they are celestrial chronometers.[26] Here the sun's contribution is outstanding: the spectacular alternations of light and darkness caused by his rising and setting provide us with "a perspicuous measure" (μέτρον ἐναργές) of all the heavenly motions.[27] But the moon's eastward circuit of the heavens also affords us with a serviceable temporal yardstick, the lunar month.[28] So does the annual circuit of the sun. The periods of the five planets too, if they had been determined, would have provided further units of time reckoning.[29] And while the like contribution of the fixed stars is not mentioned, it is acknowledged by implication when Plato speaks of all the

24. The fact that their creation is mentioned only at 40A–B, after the creation of sun, moon, and five planets (38C–39D), is an accident of the exposition; it has no chronological import, since chronology starts with the creation of the heavens ("time and the heavens were generated together," 38B6), and "the heavens" *is* the fixed stars along with the "wandering" septet.

25. I.e., their translatory "forward" (40B1) motion. They have another motion as well—axial rotation—invisible to us (40A7–B1), as must all the celestial bodies, including the ones that "wander": this is not spelled out in their case, but is undoubtedly meant to hold for them no less than for the fixed stars (that they "think always the same thoughts about the same things," *loc. cit.*, would hold in their case as well).

26. "Instruments of time": ὄργανα χρόνων, 41E5; ὄργανα χρόνου, 42D.

27. For the citation of the passage in which this is said, and for a problem that passage raises, see Appendix, section C.

28. I.e., the lunar synodic month: the interval from new moon to new moon, in which "the moon traverses its own circuit and overtakes the sun" (39C3–4).

29. See 39C5–D2, quoted, with comment on its import, in the Appendix, section D.

stars—not just the "wandering" septet—as "instruments of
time,"[30] and again when he says that "time and the heavens
were generated together" (38B6), for the fixed stars make
up the overwhelming majority of the constituents of the
heavens.[31]

Where in all this, the reader may wonder, is the coherent
scheme organizing scientifically ascertained facts that I prom-
ised a few pages back? I shall come to this, but not before
attending to two preliminary tasks: (A) I must make clear
what I understand by "scientifically ascertained facts"; (B) I
must establish that facts of this sort were now available in
Greek astronomy and that Plato himself was taking full advantage of their availability. Task A can be quickly disposed of.
By "scientifically ascertained facts" I understand facts satisfy-
ing three basic requirements:

1. They are established by observation or by inference from
it: they are derived, directly or indirectly, by the use of the
senses.

2. They have theoretical significance: they provide answers
to questions posed by theory.

3. They are shareable and corrigible: they are the common
property of qualified investigators who are aware of possible
sources of observational error and are in a position to repeat
or vary the observation to eliminate or reduce suspected error. Task B will take longer. What I would like to do here is
to give the reader some sense of the progress which Greek
astronomy had already made as an observational science and
show that Plato was well abreast of this progress by the time
he came to write the *Timaeus*. To document this progress in
any detail is not possible. The record is too fragmentary, and
often vague and unreliable as well. But even so, meager
though it is, the record yields matter enough to document
essential points.

We happen to know that in the thirties of the fifth century

---

30. See n. 26 above for the references. There can be no doubt here that
all the stars are meant: souls "equal in number" (ἰσαρίθμους) to the stars
have been created, and each soul gets one star for its habitation.

31. As Plato uses the term in the *Timaeus*, "the heavens" (οὐρανός) is nothing but the stars which make it up.

B.C.—two thirds of a century or so before the writing of the
Timaeus—two Athenian astronomers, Euctemon and Meton,³²
had made observations³³ which disclosed the inequality of the
astronomical seasons. Their figures were as follows:
From summer solstice to autumnal equinox, 90 days.
From autumnal equinox to winter solstice, 90 days.
From winter solstice to vernal equinox, 92 days.
From vernal equinox to summer solstice, 93 days.³⁴
32. Otherwise known for their contributions to the reform of the calendar
(for references see Heath, *Aristarchus of Samos*, p. 293). Meton was the author
of a "Great Year" of nineteen years, named after him, which aligned the
lunar month with the solar year by intercalating seven months in the course
of the nineteen-year period and by using months varying in length between
twenty-nine and thirty days. "This would give a mean lunar month less than
two minutes too long" (Dicks, *Early Greek Astronomy*, p. 88). A similar interca-
lation cycle (seven intercalations in nineteen years) comes also into use in
Babylonia during the fifth century (O. Neugebauer, *The Exact Sciences in Antiquity*, p. 97), but there is no evidence that Meton derived his own cycle
thence. According to Theophrastus (*de Sign.* 4) he derived it from a certain
Phaenus, an astronomer who was a resident alien in Athens.
33. By rare good luck we have an exact date for one of these: Ptolemy,
writing half a millennium later, refers to their observation of a summer solstice in 432 B.C. (*Syntaxis Mathematica* ["The Almagest"] 3. 1 [Claudii Ptole-
maei *Opera Omnia*, J. L. Heiberg, ed., vol. 1, part 1, p. 205]).
34. These figures are preserved in the papyrus entitled "Ars Eudoxi," Column 23 (F. Blass, ed., Eudoxi *Ars Astronomica*, p. 25). The papyrus mentions
only Euctemon. But Eudemus (in the citation which is to follow) credits the
discovery of the inequality of the seasons to both Euctemon and Meton. They
are credited jointly with the astronomical observation by Ptolemy in the locus
cited in n. 33 above (though on p. 206 of the same work he mentions only
Euctemon in connection with the same observation); and in another work
(*Phaseis* [*Opera Omnia*, J. L. Heiberg, ed., vol. 2, p. 67]) Ptolemy attributes
jointly to the pair meteorological observations (ἐπισημασίαι: observations
which were believed to allow weather prognostications) "in Athens and in
the Cyclades and in Macedonia and in Thrace." So it is safe to credit them
both with the above figures. These are to constitute an important datum
for subsequent astronomical theorizing. Thus we know on the authority of
Eudemus that one of the important modifications of the Eudoxian theory
was made by Callippus in order to bring the system of homocentric spheres
into line with these figures:

> Eudemus stated briefly what are the phenomena for the sake of which Callippus
> thought those additional spheres [to which Aristotle refers in *Metaph.* 1073B32ff.]
> were required. For, he [Eudemus] says, that Callippus asserted that, if the periods
> between the solstices and the equinoxes differed by as much as Euctemon and
> Meton thought they did, three spheres [those which Eudoxus had thought would
> suffice to account for the motions of the sun and the moon, three spheres in each
> case] would not suffice to save the phenomena, because of this observed irregular-
> ity. [Eudemus [fr. 149 Wehrli] *ap.* Simplicius, *in de Caelo* 497.17–22]

What is remarkable about these findings is not their accuracy. As Heath (*Aristarchus of Samos*, pp. 215–16) points out, compared with modern figures they show errors ranging from 1.23 to 2.01 days, while the error in the corresponding figures reached a century later by Callippus (92, 89, 90, 91 days, respectively)[35] ranges only from 0.08 to 0.44 days. The truly memorable thing here is that Euctemon and Meton were prepared to allow observation to supersede the assumptions that the two intersolstitial intervals are strictly equal and that both equinoxes fall at their exact midpoints—assumptions so very plausible in themselves and so seductive in their own time and place, given the Greek obsession with symmetry. Banking on these assumptions, Euctemon and Meton might have spared themselves the trouble of reaching by observation figures for equinoxes and winter solstices: they might have got these by simply counting days between successive summer solstices and dividing by four—which is what Babylonian astronomers were doing down to the end of the fourth century B.C.[36] That they chose instead the harder, observational way, and that they stuck by its results when it turned up inequalities in lieu of the expected equalities, is striking testimony to the observational orientation to which practicing Greek astronomers were now committed.

And what is involved here is not simple, but theory-charged, observation which, as Dicks has claimed, "presupposes, at the very least, the theory of a spherical earth at the center of the celestial sphere" (*Early Greek Astronomy*, p. 88).[37] The argument for this claim need not be premised on the assumption that Euctemon's observational procedures were

35. Also preserved in the papyrus cited in the preceding note.
36. Neugebauer (*The Exact Sciences in Antiquity*, p. 102), describing Babylonian practice: "It is the summer solstices which are systematically computed, whereas the equinoxes and the winter solstices are simply placed at equal intervals."
37. A very substantial claim, for the question whether the earth is flat or spherical remains in dispute among philosophical astronomers down to the end of the fifth century and even beyond: Plato so represents it in the *Phaedo* (97D–E); and we know from Aristotle (*de Caelo* 294B13–14 [= DK 13A20]) that not only Anaxagoras (fifth century) but even Democritus (whose lifetime extends well into the fourth) had clung to the view that the earth is flat.

the same as those in use in subsequent, more highly developed, stages of Greek astronomy.[38] This assumption would be hard to justify in the face of the much higher accuracy in Callippus' figures, which suggests drastic improvements in observational methods in the course of the next hundred years. The substance of Dicks's claim can be supported by arguments which bypass this risky assumption: Euctemon's and Meton's acceptance of the sphericity of the earth may be inferred with a measure of probability from their peregrinations north and south in carrying out their observations:[39] they are not likely to have done this if they had been unable to handle latitude;[40] and the concept of latitude certainly presupposes a spherical earth.[41] Moreover, there can be little doubt that the obliquity of the ecliptic was known to their contemporary, Oenopides.[42] It therefore

38. In "Solstices, Equinoxes, and the Presocratics" (p. 32) Dicks had listed a complex set of theoretical assumptions (including the celestial and terrestrial spheres), claiming that these were presupposed in the earliest equinoctial calculations by Greek astronomers, buttressing up the claim with the remark that this "is clear from the methods used by Hipparchus and Ptolemy."

39. See n. 34 above.

40. Their meteorological observations are dated by risings and settings of fixed stars which would be affected by variations of latitude.

41. A powerful argument for the sphericity of the earth in Aristotle is the observed variation of celestial phenomena that results from changes in the observer's latitude: "Certain stars visible in Egypt and in Cyprus are not visible in northern locations, and some stars which never set in northern areas do set in the aforenamed places" (de Caelo 298A3–6). This is a standard argument in later authors; see, e.g., Theo Smyrnaeus, pp. 191–92.

42. From the testimonia in DK 41A7 we learn the following: Oenipides claimed "as his own invention" the discovery of the ecliptic (Aetius, de Placitis Reliquiae [hereafter to be cited by author's name only] 2.12.3). He is said (Diodorus Siculus, Bibliotheca Historica 1.98.2) "to have consorted with priests and astronomers [in Egypt] and learned among other things that the circular path of the sun is oblique and that [the direction of] its motion is opposite to that of the other stars." Eudemus ([fr. 145 Wehrli] ap. Theo Smyrnaeus, p. 148 [citing Dercyllides as his authority]) relates that Oenopides "was the first to discover the obliquity of the zodiac"—the unamended text reads "cincture (διάζωσιν) of the zodiac," but Diels's emendation λόξωσιν for διάζωσιν has been widely accepted. To smooth out the inconsistencies we need only suppose that Oenopides learned of the obliquity of the zodiac from Egyptian sources and then discovered for himself the obliquity of the ecliptic as the trajectory of the sun's annual eastward course through the zodiac. From Oenopides' interest in the astronomical bearing of geometrical constructions

would not be unreasonable to assume that they too had as-
similated this discovery[43] and its implied conceptual machin-
ery: the heavenly sphere with its equator as a great circle at
right angles to its north-south axis; the tropics as two circles
on the sphere, parallel to the equator and equidistant from
it; and the ecliptic as another great circle on the sphere, intersecting the equator at a sharp angle and touching the tropics
at its northern and southern extremities.

Even if we had missed this information about Euctemon
and Meton, we could still have inferred with confidence that
well before Plato's time—by the last third of the fifth century
at the latest—the role of observation had changed drastically
in Greek astronomy since the days when rational inquiry into
the heavens had begun in Miletus as a branch of *physiologia*.
We could have inferred this directly from the intervening
changes in the reported content of astronomical doctrine.
To mark the contrast let me refer again to that grandiose
model of the cosmos to which I alluded in the preceding
chapter:[44] Anaximander's. Here the sun, it will be recalled, is
an enormous ring or hoop of fire at the periphery of our

(attested by Proclus, *Commentarii in Euclidem*, G. Friedlein, ed., p. 283 [here-
after to be cited by author's name only], English translation, *Proclus: A Commentary on the First Book of Euclid's Elements*, G. R. Morrow, p. 220), von Fritz
(article on Oenopides in Pauly-Wissowa, *Real-Enzyklopädie der klassischen Al-
tertumswissenschaft* 17:2258-71 at 2260-62, and *Grundprobleme der Geschichte
der antiken Wissenschaft*, p. 139) has inferred with considerable plausibility that
it was Oenopides who fixed the value of the obliquity of the ecliptic at that of
the angle subtended by the side of a regular fifteen-sided polygon (cf. n. 20
above).

Ascriptions of the discovery of the obliquity of the zodiac to earlier figures—even to Anaximander (DK 12A5 and 22)—are not trustworthy in my
opinion; but see Kahn's arguments for what he calls "the empirical discovery
of the zodiac" ("On Early Greek Astronomy," pp. 103-6).

43. Proclus (p. 66) speaks of Oenopides as "a little younger" than Anaxagoras (who was born around 500 B.C.). If he were, say, fifteen years younger,
it would be reasonable to place his astronomical discoveries in the forties of
the fifth century (or, at any rate, not much later), hence likely to be known
in Athens by the time Euctemon and Meton were making their observations
(cf. n. 33 above). Their interest in Oenopides' astronomical ideas would be
heightened by the fact that they shared his preoccupation with the reform
of the calendar (he is the author of a "Great Year" [cf. n. 32 above] of fifty-
nine years; for the testimonia see DK 41A9).

44. Pp. 7, 20, above.

world, invisible save for an orifice from which the fire escapes; the moon is another fire-ring, similar in design, but smaller, its diameter two thirds that of the sun; this is followed in turn, after another interval of vacancy, by the countless multitude of star-rings whose visible orifices are tiny and whose diameters are half that of the moon, a third that of the sun.[45] A mere glance at this construct will show that it is a work of the imagination whose departure from the inherited world-view could not have been justified by a single scientifically ascertained fact meeting the conditions I laid down above (pp. 7, 20). Anaximander is said by a late author to have discovered the gnomon—that simple, but enormously useful, instrument consisting of a rod fixed upright on a flat base on which seasonal variations in the length of the sun's shadow may be measured.[46] The veracity of the report is dubious.[47] But suppose it were conceded: suppose Anaxi-

---

45. For the textual data on which this account is based see, e.g., Kirk in Kirk and Raven, *The Presocratic Philosophers*, pp. 135–37: the main doxographical texts, with translation and commentary; fuller treatment in Kahn, *Anaximander*, pp. 58–63. The picture is conjectural to a degree, the figures for the stars wholly so, being due to Paul Tannery's ingenious inferences (*Pour l'Histoire de la Science Hellène*, pp. 90ff.), and their soundness has been vigorously disputed (Dicks, "Solstices, Equinoxes," p. 36; but see D. O'Brien, "Derived Light and Eclipses in the Fifth Century," pp. 114–21, and further exchanges: Dicks, "On Anaximander's Figures," p. 120; O'Brien, "Anaximander and Dr. Dicks," p. 198); but they have been accepted by the overwhelming majority of historians.

46. Diogenes Laertius, *Vitae Philosophorum* (hereafter to be cited by author's name only) 2.2 (in DK 12A1): "He was the first to discover the gnomon and he set it up at the Skiathera in Sparta, as is stated by Favorinus in his *Universal History*." And Herodotus states that "the Greeks" learned "the *polos* and the gnomon and the twelve parts of the day from the Babylonians" (2.109.3). Later testimonia simply repeat or conflate these two reports, the second of which makes no mention of Anaximander, while the historical value of the first is uncertain: Favorinus is a polymath of the first half of the second century A.D., whose voluminous *Universal History* (in twenty-four books) is a miscellany of uneven historical credibility. Along with undoubted facts it includes patent misinformation (as when it reports, [according to Diogenes Laertius, 8.83] that Alcmaeon was the first to compose a physical treatise; we know that Anaximander had composed one two generations earlier).

47. Taken as a whole it would be rendered suspect by its tale that Anaximander set up his gnomon in Sparta, of all places (notoriously unfriendly to philosophical and scientific imports) and its representation as equipped to

mander did have a gnomon and used it to his heart's content. Nothing he could have learned with its help could have yielded the slightest observational support for the sun-moon-stars-earth order in his model.[48] Conversely, had he troubled to observe the starlit sky over extended periods of time he would have been virtually certain to pick up facts irreconcilable with that order, for example, facts of celestial occultation.[49] These are the very ones to which Aristotle refers as observational evidence for the moon's being nearer the earth than are the planets or fixed stars:

> For we have seen the moon at the half-full approaching on its own darkened side the star of Mars which is thereupon screened from our view to reappear on the moon's bright and shiny side. About the other stars too similar accounts have been given by Egyptians and Babylonians who had observed them for very many years in the past and from whom we have received many trustworthy reports about each.[50]

---

indicate "solstices and equinoxes" (unlikely at this earliest phase of Greek astronomy: Dicks, "Solstices, Equinoxes," pp. 29–30), even if Favorinus had a more consistent record of historical credibility. However, when all the objections to his testimony have been given full weight, we cannot be sure that there is no fire behind his smoke. For all we know to the contrary, Anaximander may have made use of some simple form of the gnomon.

48. For the sort of reason Anaximander might have had for hitting on this perplexing order, see Kahn's ingenious surmise (*Anaximander*, p. 90): he might have been counting on "a general tendency for fire to collect more abundantly near the periphery of the heavens" (typical theorizing in the style of Ionian *physiologia*).

49. Cf. Dreyer's remark on Anaximander's placing the stars between the earth and the moon: "This shows at once how little the celestial phenomena had been watched, as the frequent occultation of a bright star by the moon must have been unknown to Anaximander" (*A History of Astronomy from Thales to Kepler*, p. 14).

50. *De Caelo* 292A3–9. The dating of the influx from Babylonia into Greece of extensive tables of observations of planetary motions has long been in dispute. For selective references to the contending authorities see Dicks (*Early Greek Astronomy*, pp. 168ff. and notes). The *terminus ante quem* is the *Epinomis* (probably by Plato's pupil, Philip of Opus, early in the second half of the fourth century) and Aristotle's *de Caelo* and *Meteorologica* (not much later). The first two treatises allude to astronomical data of immense antiquity emanating from the Orient ("Egypt and Syria," *Epinomis* 987A; "Egyptians and Babylonians," *de Caelo* 292A7–9; "barbarians," *de Caelo* 270A5–16). In the third (*Meteor.* 343B10–11 and 28–30) Aristotle draws on astronomical observations by "Egyptians" in rebutting Democritus' theory of comets (n. 62 below). The absence of any earlier allusions to detailed observational material

Aristotle's mention of "the star of Mars" and his allusion to similar records about other planets points up another feature of Anaximander's astronomical theory symptomatic of the scarcity of scientifically ascertained data available to its author: its world-model makes no special place for planets in contradistinction to the fixed stars.[51] It is doubtful that it even took explicit notice of the existence of any celestial body which would have answered to what the Greeks understood by "planet"—stars (other than the sun and moon) distinguished from the fixed stars by their apparently erratic ("wandering" or "roaming") motion. From our (admittedly meager) sources we cannot even be sure that any star had been identified in Greece as a planet before Parmenides (first third of the fifth century B.C.). Though Venus was known and observed in Babylonia even before 1,500 B.C.,[52] there is no sign of it in Homer or Hesiod—no indication in those poems that "Dawnbringer" (Ἑωσφόρος) and "Evening Star" (Ἕσπερος) are named for the same star and that what each names is a "roamer."[53] As late as the fourth century B.C.

----

from the Orient (most particularly its absence in the *Timaeus* and the *Laws*) makes it rather unlikely that this influx had occurred much earlier than the middle of the fourth century.

51. Statements to the contrary in accounts of his astronomical theory in some histories of Greek philosophy and science are without foundation, as I explain in the Appendix, section E.

52. A. Pannekoek, *A History of Astronomy*, pp. 33–35. All five planets are known and observed in the Assyrian period (eighth to seventh century B.C.): ibid., pp. 39–41, citing texts of the seventh century B.C.

53. Cf. F. Cumont "Les noms des planètes et l'astrolatrie chez les Grecs," *Antiquité Classique*, pp. 5–6; Dicks, *Early Greek Astronomy*, pp. 32–33. That the identification had been missed is certain. What is uncertain is whether or not either of these putatively separate stars is thought of as a planet in the epic. The answer turns on whether or not we are entitled to assume that the marked differences between the motion of each and that of the fixed stars had been noticed. Bicknell, "Early Greek Knowledge of the Planets," p. 49, argues for the affirmative, since "the change of position of Venus relative to the fixed stars is conspicuous over even a fraction of the period during which it is morning or evening star." He could be right; but the fact remains that contexts in which the "Dawnbringer" is contrasted with the fixed stars (as in Hesiod, *Theogony*, vv. 381–83) make no allusion to differences in their respective patterns of motion. Bicknell has also argued (pp. 50–51) that a planet named μεσόνυξ had been identified as early as the lifetime of the poet Stesichorus (late seventh, early sixth century). But the only text on which his

there are parts of Greece where the identity is unknown: the Cretan interlocutor in Plato's *Laws* has no suspicion of it.[54] The identification is assigned to Parmenides in a doxographic text of respectable authenticity:[55] "First in the aether Parmenides places the Dawnstar (τὸν Ἐῶιον) which he thinks identical with the Evening Star; following this, the sun, under which he puts the stars in the fiery region which he calls 'heaven'" (Aetius 2.15.7 [=DK 28A40a].[56] The text does not state that Parmenides thought of the Dawnstar as a "roamer." But the fact that he makes of it a distinct "garland" (στεφάνη),[57] separated from the fixed stars by the "garland"

---

claim is based is a note in an obscure grammarian of the sixth century A.D. (μεσόνυξ: εἰς τῶν ἑπτὰ πλανητῶν παρὰ τοῖς Πυθαγορείοις ὀνομάζεται· μέμνηται Στησίχορος), and it is not clear from the wording that, if Stesichorus did indeed so speak of some star or other, the context was such as to give the grammarian (or his source) any good ground for inferring that the star Stesichorus was talking about was "one of the seven Pythagorean planets" or any sort of planet.

54. 821C: "In my own lifetime I have often observed that both the Dawnbringer and the Evening Star and certain other stars never travel on the same course but wander about in every way; and that this is how the sun and the moon behave we all know." The grammar of the speaker's allusion to the first-named stars (τόν τε Ἑωσφόρον καὶ τὸν Ἕσπερον) gives away his lack of awareness of their identity.

55. In addition to this text see also Diogenes Laertius, 9.23 (in DK 28A1): "He is believed to have been the first to state that the Evening Star and the Lightbringer are the same." Diogenes' authority being Favorinus, the report may not be worth much (cf. n. 46 above); but it does at least show that neither Diogenes nor Favorinus knew of any ascription of the discovery to an earlier thinker.

56. The use of Ἑῶιος instead of the customary Ἑωσφόρος suggests that Aetius here is using a source which reproduces Parmenidean diction, which is also suggested by the terminal clause (probably an allusion to B10, 5). The veracity of this report has been widely accepted (Heath, *Aristarchus of Samos*, p. 75; O. A. Gigon, *Der Ursprung der griechischen Philosophie*, p. 277; L. Tarán, *Parmenides*, p. 242; Dicks, *Early Greek Astronomy*, p. 51). No good reason for rejecting it has ever been given, to my knowledge; Guthrie (*History of Greek Philosophy* 2:57ff.) gives none for omitting it from the texts relating to Parmenides' cosmology.

57. As has long been recognized, these "garlands" or "bands" are a variation on Anaximander's conception of celestrial bodies as great rings or hoops. Parmenides' choice of στεφάνη for this purpose is in line with his penchant for the diction of the epic (for the references see Gigon, *Der Ursprung der griechischen Philosophie*, p. 277).

of the sun,[58] suggests that he recognizes that its own pattern of motion is very different from theirs.[59] So we should probably reckon Parmenides the first Greek to identify one of the five planets of Greek astronomy.[60]

Thereafter progress was rapid: the other four, and their approximately correct order,[61] were discovered within two generations of the identification of the first. This solid achievement of observational astronomy was won in spite of some continuing addiction to highly speculative hypotheses, projected with no apparent concern for observational controls. In that vein Anaxagoras and Democritus postulate the existence of an unspecified multitude of planets, claiming that the conjunction (σύμφασις) of such bodies is what accounts for the appearance of comets;[62] and Democritus'

58. In placing the fixed stars under the sun he appears to be following Anaximander, while breaking with him in giving the moon the lowest place in the heavens (Aetius 2.7.1 [= DK 28A37]), which would allow, as the Anaximandrian model would not, for occultations of stars by the moon.

59. Indeed he could scarcely have failed to recognize this in the course of making the observations and inferences which led him to discover the identity of the Morning and the Evening Star.

60. For argument in support of this claim on behalf of Parmenides, see Appendix, section F.

61. I.e. their correct order in radial distance of their orbits from the earth (which is Mercury–Venus–Mars–Jupiter–Saturn) except that Venus is thought to precede Mercury by most Greek astronomers (as by Plato in *Republic* 616E–617B and *Ti.* 38D) down to the second century B.C. (for the references see A. E. Taylor, *Commentary on Plato's "Timaeus,"* 192–94). The contrast with Babylonian astronomy is striking. It had got off to a much earlier start (cf. n. 52 above)—all five planets had been identified by the seventh century at the latest (i.e., at least two centuries before Parmenides)—but the discovery of their correct order was missed down to the end of the fourth century B.C.: "In the cuneiform texts of the Seleucid period the standard arrangement is

Jupiter–Venus–Mercury–Saturn–Mars.

The reason for this arrangement is unknown . . ." (Neugebauer, *The Exact Sciences*, pp. 168–69).

62. Aristotle, *Meteorologica* 342B27–29. The "wandering stars" whose concourse is supposed to explain the appearance of comets could not have been thought members of the quintet which was to become canonical in Greek astronomy: no pair from this group could have been plausibly thought to stay together at some conjunction long enough to account for the appearance of a comet and to answer to the supposed fact (adduced, according to Aristotle, by Democritus in "contentious defense" of his theory) that "stars

Pythagorean contemporary, Philolaus, turns the earth itself
into a star revolving about an invisible fire at the center of
the universe and interpolates another star, a mysterious
"counter-earth," in between.[63] Yet side by side with these
fanciful flights,[64] observationally grounded theory was ad-
vancing apace. The true quintet of planets is known to Philolaus[65] and possibly also to Democritus.[66] In any case, it

---

[presumably, the ones in the conjunction] have been seen to appear at the
dissolution of comets" (*Meteor.* 343B27-28): one can only suspect that Democritus was going by hearsay; as Aristotle implies, the supposed phenomenon could not have been grounded on observation.

63. In Aristotle these remarkable doctrines are ascribed to "those in Italy
who are called 'Pythagoreans'" (*de Caelo* 293A20-25). In doxographic reports they are credited by name to Philolaus (Aetius 2.7.7 and 3.11.3 [= DK
44A16, 17]). Aristotle (*de Caelo* 293A25-28) denounces the anti-empirical
temper of these doctrines. The doxographic ascription to Philolaus has been
disputed. But see Burkert (*Lore and Science*, pp. 337ff). Agreeing with him, I
hold that none of the objections are conclusive and that the balance of the
evidence favors the ascription. That Aristotle refrains from mentioning Philolaus' name is not at all decisive: his reticence is in line with his general policy
of discussing the views of Pythagoreans in block-terms, declining to refer to
individual members of the school.

64. Quite apart from the mysterious counter-earth, even the notion of the
earth's orbiting around a central point of the planetary system could hardly
have been supported by facts observed, or observable, at this time. Our present knowledge that this notion represents physical truth does not, of course,
improve its empirical credentials in its own time. In the light of facts then
known the case for an immobile earth was overwhelming (cf. T. S. Kuhn,
The Copernican Revolution, pp. 84-87). How Philolaus could have thought
that his own theory accounts for the same evidence remains unclear; for the
difficulties such a theory would have had to face see the perplexed discussion
in Heath, *Aristarchus of Samos*, pp. 101ff.

65. Five planets are ascribed to him explicitly in the first of the doxographic reports cited in n. 63 above. His adherence to the same number may
also be inferred from the fact that it would take just that many planets (along
with central fire, counter-earth, moon, sun, and sphere of fixed stars) to
bring up the number of heavenly bodies to ten, which is reckoned the "perfect" number in Pythagorean lore; according to Aristotle (*Metaph.* 986A6-
12) the very reason for which the Pythagoreans postulated a counter-earth
was to get this "perfect" number. That he had hit on the correct order may
be inferred from Eudemus' report (cited in Appendix, section E) that the
Pythagoreans discovered the positional order of the "wandering" stars. Eudemus' "Pythagoreans," like Aristotle's, were almost certainly Philolaus (see
n. 63 above) who is the only fifth-century Pythagorean known to have produced an important astronomical theory.

66. For Democritus' probable views about the planets see Appendix, section G.

must have been well established by the time Plato came to write the *Republic* (probably late in the second decade of the fourth century),[67] for he builds it into the world-model of the myth of Er: there the earth is at the center of the cosmos, the fixed stars are at its periphery, and in between these two extremes come seven stars which could only be moon, sun, and five planets. The last are identified by color, not by name, but there can be no doubt as to which is which and that they are placed in the right order.[68]

I have taken the time to trace the discovery of the planets because a finding of this nature establishes conclusively the coming of age of Greek astronomy[69]—its transformation from the speculative exercise it had been throughout its origins in Milesian *physiologia* and for at least a generation thereafter[70] into a discipline where theory, continuing to be as boldly imaginative as ever, had now entered into produc-

67. Though much in Plato's biography is uncertain and controversial, there is fairly wide agreement among scholars that the *Republic* was composed at a date close to that of his first journey to Syracuse, which is known to have occurred when he was "nearly forty" (Epistle VII, 324A6), i.e., at, or just before, 387 B.C.

68. *Republic* 616B–617D. See Heath, *Aristarchus of Samos*, pp. 155–58.

69. More so than does the more sensational discovery of the cause of eclipses: this is a triumph of inventive inference rather than of observationally controlled theory. Thus in the case of the lunar eclipse once one has come to think of (1) the sun and the moon as bodies orbiting about the earth, (2) the moon as a dark body deriving its illumination from the sun, and (3) the earth casting a shadow, then, just by putting these three things together *without any further recourse to observation*, one could have hit on the idea that if the moon were to run into the earth's shadow it would be certain to be eclipsed. How little the true theory of eclipses, as held by Anaxagoras and Democritus (see O'Brien, "Derived light and Eclipses," pp. 114ff.), owed to patient interrogation of the phenomena may be judged from the following consideration: had they paid closer attention to the apparent shapes which the full moon assumes in the course of an eclipse, they would have found there evidence irreconcilable with the notion of a flat earth to which both still adhered (see n. 37 above). As Aristotle points out (*de Caelo* 297B23–30), the eclipsed moon exhibits a variety of figures which, taken as a whole, would result if and only if the moon had moved into the shadow of a *spherical* body.

70. Dicks's remark (*Early Greek Astronomy*, p. 60) about the astronomical theories of "the earlier Presocratics," that "they are the dream children of the speculative thinker in his study intoxicated by the novelty and daring of the new intellectual atmosphere," would be entirely correct of the Milesians, Heraclitus, and Alcmaeon. His further remark, "They were not primarily scientists, much less astronomers, and observation of actual celestial

tive teamwork with observation, its hypotheses now being informed by a growing body of scientifically established data.
To say that those five planets were discovered is *ipso facto* to
refer to a multitude of observations brought into systematic
unity by theory. It is to assert that at particular places in the
sky at particular hours of the night in particular periods of
the year shining objects of a certain order of brilliance had
been seen and would be always there to be seen by anyone
who cares to look; and that these five sets of visibilia represent as many stars, each set answering to the same star making
appearances widely separated in time and space—as widely, for
example, as would be those of Mercury in a given year,
brightly visible as morning star during September and October, and then dropping out of sight to reappear once again
months later as evening star in the western horizon. Even if
we knew nothing more than that the existence of the five
planets had been discovered by Plato's time, we would know
that in the area of astronomy the Greeks had now succeeded
where they had failed in every other major area of empirical
investigation:[71] they had managed to break out of the style

phenomena seems to have played a relatively minor role in their thinking,"
would also be true of those I have just named and of Parmenides, Empedocles, Leucippus, and Anaxagoras as well. But he does the Presocratics less
than justice when he dates to "the latter part of the fifth century" the "more
empirical attitude which was prepared to take into account the facts of actual
observation." If "empirical attitude" is to be glossed by readiness "to take
into account the facts of actual observation," then all the Presocratics would
qualify: fierce rationalists though they all were, to say that "they were prepared
to take into account" whatever facts of observation happened to come their
way would be true of every last one of them. The trouble with them is that
they failed to see how sadly they were limited by the paucity of available facts
and how essential it would be to so design the structure of their theories as to
facilitate confrontation with the relevant facts just as soon as these became
available. See the remarks on pp. 51–53 of my review of Cornford's *Principium
Sapientiae* in Furley and Allen, *Studies in Presocratic Philosophy*, pp. 42–55.

    71. Some distinguished historians of Greek philosophy and science have
looked to medicine for the Greek paradigm of empirically oriented science.
In my review of Cornford (cited in the preceding note) I have tried to explain
why this is a mistake. Certainly the subject-matter of the medical writers is
only too plainly—and grossly—empirical. But their theoretical treatment of it
suffers from the same flaw which afflicts the *physiologoi*: the conceptual design of their theories is not made with a view to confirmation or disconfirmation by empirical data.

of free-wheeling speculation so characteristic of *physiologia* and to discover another mode of natural inquiry where theorizing was to be controlled by factual subject matter meeting the requirements of scientific observation I laid down above. The creation story of the *Timaeus*, despite its allegorical tincture, attests Plato's assimilation of the results obtained by this science in which theory and practice were now successfully interacting. His story shows how much more than the identity and order of the five planets Plato had learned from the astronomers. In particular, he had come to know

(1) that the motion of the planets, though sharing the diurnal westward movement of the fixed stars, is subject also to a much slower inverse eastward movement "aslant" to the celestial equator;[72]

(2) that there are differences in the size of their orbits, in the periods of their revolutions, and in their angular velocities relative to the fixed stars: more particularly, that Venus and Mars have periods of revolution which are equal to each other's and to the sun's,[73] and that the other planets have larger orbits, longer periods and, by his reckoning, lower speeds;[74]

---

72. See pp. 32–34 above.

73. This particular—that the (sidereal) periods of Venus and Mercury are (on the average) a solar year, which is true for a geocentric hypothesis and is generally thought true by Greek astronomers (see Dicks, *Early Greek Astronomy*, p. 186, and nn. 174, 345)—is mentioned both in the *Republic* (617A8–B1: the fifth, sixth, and seventh stars [Mercury, Venus, the sun] proceed ἅμα ἀλλήλοις, "at the same pace") and in the *Ti*. (38D1–3: the Dawnstar and the star of Hermes "run a course which is on a par with the sun's in respect of speed" [τάχει ἰσόδρομον ἡλίῳ κύκλον ἰόντας]). (Cf. also the reference to Hermes as "keeping pace (ὁμόδρομος)" with both the sun and with Venus in the *Epinomis* [987B].)

74. This too is implied both in the *Republic* (617A–B: the fastest revolution is that of the "eighth" circle, the moon's; next are those of Venus, Mercury, sun [see n. 73 above]; increasingly slower are those of Mars, Jupiter, Saturn) and in the *Ti*. (36D5–6: "three of them [are ordered by the Demiurge to revolve] at equal speeds while the [other] four at speeds which differ from one another and from that of the three,") as well as in the *Laws* (822A8–C5: people believe that the swiftest of these bodies is the slowest—a shocking, ungodly, error). Holding that the speeds become increasingly smaller as the orbits and periods become increasingly larger (*Ti*. 39A2–3), Plato is reckoning as the proper speed of each planet the one it derives from the movement of the

(3) that planetary orbits exhibit peculiarities which he describes as "back-circlings"[75] and "advances" of planets relative to one another—which could only be a reference to the
phenomena of planetary retrogradation, that is, to the fact
that the planets in their eastward progress through the constellations of the Zodiac do not always move steadily forward,
but occasionally start slowing down, appear to come to a halt,
and then actually loop backward ("back-circling") for a time
before resuming their "advance" once again.[76]
The passage in which the reference to (3) occurs is worth
quoting in full. It is the terminal remark in the account of the
creation of the stars:

> To describe the patterned movements [literally "choreography"
> (χορείας)] of these gods, their juxtapositions (παραβολάς), and
> the back-circlings and advances of their orbits on themselves (τὰς
> τῶν κύκλων πρὸς ἑαυτοὺς ἐπανακυκλήσεις καὶ προχωρήσεις);
> to tell which of the gods come into line with one another at their
> conjunctions (ἐν ταῖς συνάψεσιν) and which of them are in oppo-
> sition (καταντικρύ), and in what order and at which times they
> come in front of others, so that some come to be screened from
> our view to reappear once again, thereby bringing terrors and
> portents of things to come to those who cannot reason—to tell
> all this without [using] visible models would be labor spent in
> vain. [40C4–D3]

Plato's language here is saturated with the terms of observational astronomy:[77] "Juxtaposition" occurs when heavenly
bodies rise and set in close proximity; "conjunction" when

Different. He does not state, and does not claim to know, how much larger
are the orbits and periods of Mars, Jupiter, and Saturn than those of Venus,
Mercury, and sun. When he insists that they all "move according to ratio"
(*Ti.* 36D6–7, ἐν λόγῳ φερομένους, and cf. 39D2–7), which could be only
an act of faith on his part (a deduction from their teleological function as
"instruments of time" [ὄργανα χρόνων, 41E]), it is quite possible that he has
no figures for all of these ratios (cf. Appendix, section C). Thus the period
of Saturn, nearly thirty years, would have required more than a generation
to establish, and it is possible that it had not yet been fixed to the thirty-year
period accepted by Eudoxus according to what is said in the account of his
planetary theory in Simplicius (*in de Caelo* 493.11–497.5) at 495.28.
   75. For the Greek term I translate "back-circlings" above, see Appendix,
section H.
   76. More on retrogradation in Appendix, section I.
   77. Explained by Proclus (*in Timaeum* 284C).

they have the same celestial longitude; "opposition" when their celestial longitude differs by 180°; "screening" is occultation, but in this context it refers particularly to solar and lunar eclipses; and, as I have just intimated, "back-circlings" and "advances" are incidents of retrogradation.

So there can be no doubt that by the time Plato came to compose the *Timaeus* he was *au courant* with observational astronomy. This is not to suggest that he had himself become a practitioner of the discipline[78]—only that he had come abreast of what the discipline had discovered. What bearing then could his own philosophical tale about the stars have on that kind of work? What would it have to say to the working astronomers that would make a difference to their own observationally controlled theory? Let us look again at this tale, beginning with two of its main theses:

> Thesis *A*. The stars are gods and their motions are psychokinetic.[79]
> Thesis *B*. Stellar motions are circular.[80]

Now these two theses are not only logically distinct—neither, taken by itself, would entail the other—but belong to radically different areas of inquiry. *A* belongs to theology and speculative metaphysics, *B* to natural science. Yet Plato undertakes to deduce *B* from *A*. He does so by compounding *A* with two further theses of the same ilk:

> Thesis *C*. The souls of the star-gods are perfectly rational.
> Thesis *D*. All perfectly rational motion is circular.

C comes from Plato's theology, which makes goodness an essential attribute of divinity,[81] in conjunction with his

---

78. Plato never makes such a claim, nor is it ever made responsibly on his behalf in classical antiquity.

79. This is the doctrine of the *Timaeus*, reported earlier (see p. 31 above). It is reasserted in the *Laws* (898D–899D). The three possibilities mooted here (898E7–899A4) envisage complications that are foreign to the simple astral theology of the *Timaeus*, but all three are declared (899A7–9) to warrant belief in thesis *A*.

80. They are all circular (κύκλοι) in the *Timaeus* (36Cff.). The circularity of celestial motions is reaffirmed in the *Laws* (898A3–6).

81. *Republic* 379Bff.

axiomatic assumption that goodness entails rationality. So
when the Demiurge populates the sky with a whole new breed
of gods whose natures, owing nothing to popular myth or
ritual, can be refashioned freely by Plato's philosophical imagination, it is inevitable that they should all be equipped with
intelligence of imperturbable rationality.[82] As for D, we have
met it already. Earlier in this chapter I cited the description
of circular motion in the *Timaeus* as "the one most befitting
reason and intelligence" (34A). The same thing is expressed
more elaborately in the *Laws*: rotary motion—described as the
one "which moves invariantly and uniformly about the same
things and in [the same] relation to the same things according
to one rule and order"[83]—is "of all motions the one which
has in every way the nearest possible kinship and similarity to
the revolution of intelligence" (898A3–6).[84]

Given theses *A*, *C*, and *D*, thesis *B* follows: if the motions
of the stars are caused by their souls, if their souls are rational, and if rational motion is circular, then all their motions
must be circular. So if Plato's metaphysical theology is true,
the grand astronomical claim made in *B* would have to be true.
Well, is it? So far as the fixed stars go, the answer for Greek
astronomy is only too obviously, "Yes." Plotting successive
positions of any circumpolar star over a twenty-four hour interval we would see at once that they fit precisely a circular

82. They are totally exempt from the perturbations of our own reason, for
these are due entirely to the "influx" and "efflux" to which our body is
subject (*Ti.* 43A; and cf. the citation from *Phaedo* 66A in n. 85 below), and
nothing of that sort afflicts their body, which is a perfectly self-contained
sphere.

83. Ideally this would be rotation *in situ*. An allowable approximation is
rotary motion about a fixed center: there translation in space maintains constancy of distance from the same stationary point, which could never happen
in any form of rectilinear motion.

84. For Plato invariance is the highest ("most divine") condition a temporal existent can achieve: "to be everlastingly in an invariant and uniform state
and to be self-identical pertains only to the most divine things" (*Politicus*
269D). The closer something comes to this state, the more nearly it escapes
the mutability which is the curse of temporal existence and approaches the
condition of the eternal Ideas, on whose absolute invariance Plato dwells often
(*Phaedo* 78C6; *Republic* 479A1–3 and E7–8; *Ti.* 29A1; *Philebus* 56C3–4).

locus. And this would remain true no matter how close those successive positions were made to be. An easy extrapolation would show that the same thing must be true of all fixed stars, including those whose diurnal revolutions are not observable in full or not observable at all from northern latitudes in some parts of the year. So far then the astronomical implicate of theses *A*, *C*, and *D* coincides exactly with a familiar, widely accepted, inference from long-established phenomena.

Not so in the case of the other stars which have "wandering" motions. To all appearance these movements do not fit the perfectly circular pattern. That is the very reason why they had come to be thought of as "wandering" in the first place. What is Plato to do at this juncture? Is he to say, 'Sensory appearances be damned if they belie a proposition which follows with deductive certainty from the rational truths of my metaphysics'? This is what might have been expected of him in view of the attitude to sense-experiences he seemed to be taking decades earlier, in the *Phaedo*: there he had assailed the senses as delusive and insisted that the philosopher should distrust them on principle and shun them so far as possible in his search for truth.[85] This downgrading of the epistemic value of sense-experience might easily have turned Plato against the methodology which had transformed the study of the heavens from inspired guesswork into empirically controlled inquiry. But as I tried to show above Plato managed to escape such a tragic alienation from the science of his time. His practice, if not his theory, shows that he would realize the folly of any theorizing about the heavens which would ignore or deny facts ascertainable by observation. His task then was to reconcile the *a priori* conviction that all celestial motion is rotary with the empirical facts concerning the "wandering" motions of seven stars. In the *Timaeus* we see

85. "For it is clear that the mind will be deceived when it attempts to investigate anything in partnership with the body" (65B). And cf. 65E6–66A6, which concludes: ". . . getting rid so far as possible of eyes and ears and so to speak of the entire body as perturbing the soul and not allowing it to come into the possession of truth and wisdom when it takes the body into partnership."

him projecting a theory designed to effect this reconciliation: he hypothesizes that the motions of sun, moon, and planets are in every case compositions of *un*wandering circular motions proceeding in different planes in different directions at different velocities—the westward movement of the Same in the plane of the celestial equator, compounded with the much slower eastward movement of the Different in or near the plane of the ecliptic.

The conceptual kernel of this hypothesis—its "central insight," if I may so call it for purposes of later reference—is that the composition of postulated regular circular motions may account for irregular phenomenal motions. Plato has hit here on a profoundly original and fertile notion—the grand heuristic canon of Greek astronomical theory for half a millennium to come.[86] More on this presently. But before proceeding to the long-range import of that central insight, let me say more on its successful application to solar motion in the *Timaeus*. The resulting theory would have been of special interest to the working astronomer, because it generated a multitude of inferences concerning the sun's expected

86. The fertility of Plato's central insight has been well recognized, e.g.: P. Duhem, "Si nous voulons trouver la source de la tradition dont nous pretendons suivre le cours [Eudoxus, Hipparchus, Ptolemy], il nous faut remonter à Platon" (*Sōzein ta phainomena*, p. 3); and E. J. Dijksterhuis, "He [Plato] sets before astronomers the methodological problem which under the name of the Platonic axiom was to dominate theoretical astronomy for twenty centuries: to detect in the confused irregularity of the motions of the planets the ideal mathematical system of uniform circular motion . . ." (*The Mechanization of the World Picture*, p. 15).

The same recognition has not been given to the originality of Plato's insight. In antiquity this was obscured by the delusion that the insight had been anticipated by Pythagoras (for references, and for the refutation of this crude anachronism, see Burkert, *Lore and Science*, pp. 325–29). In modern histories of Greek astronomy the composition of the two motions of the Same and the Different does provide the basis of the explanation of the theory of the "spiral twist" (*Ti.* 39A–B, to be discussed directly in the text above); and the novelty of this theory has been acknowledged—so Dicks: "Recognition of these spiral courses of the planets argues no small astronomical insight on Plato's part, as there is no evidence that any such notion was entertained before him" (*Early Greek Astronomy*, p. 129). But I cannot recall a just appreciation of the novelty of the more general conception—the "central insight" above—by means of which Plato had reached the conclusion that solar, lunar, and planetary orbits were all spirals.

behavior—inferences which could be used to institute new observations that might confirm or disconfirm the theory. Let me spell out one set of such inferences, the one associated with the conclusion that the composite motion of the sun is a spiral. That Plato himself has drawn this conclusion comes out clearly when he remarks that "the revolution of the Same gives to all the circles [those of sun, moon, and planets] *a spiral twist, because they move simultaneously in a twofold way in contrary directions* (στρέφουσα ἕλικα διὰ τὸ διχῇ κατὰ τὰ ἐναντία ἅμα προϊέναι)" (39A6–B1).[87] He does not tell us how he passes here from premise to conclusion. But it is not hard to figure out the steps in the case of solar motion. Here they are (see Fig. 1) as filled out by historians glossing this passage;[88] every step in the reconstruction of the reasoning keeps well within the limits of elementary spherical geometry which would be perfectly familiar to Plato.

Figure 1 represents an imaginary sphere concentric with the celestial sphere, but much smaller (its radius is the distance from the center of the earth to the center of the sun). *NS* is the polar axis. *CQDP* is a great circle on our imaginary sphere in the plane of the celestial equator. If the sun's motion were derived exclusively from the movement of the Same it would constitute a diurnal westward revolution about the circle *CQDP*. *EPBQ* is the ecliptic, a second great circle on our sphere, the trajectory of the sun's eastward revolution through the zodiac. The lines *AB* and *EF* are stand-ins for circles on our imaginary sphere in the planes of the northern and southern tropics. (Their representation by lines is schematic: to draw them as circles would clutter up the diagram too much.)

Let the center of the sun be at *Q* at a given moment when the planes of the celestial equator and the ecliptic are intersecting at *QP*. The movement of the Same is carrying the sun westward along the circle *CQDP*. Concurrently the move-

---

87. It has been argued that this doctrine had been anticipated in *Ti.* 36D4–5. See Appendix, section J.

88. T. H. Martin, *Études sur le Timée de Platon*, 2:76. A fuller account in P. Duhem, *Le Système du Monde*, pp. 54–57.

ment of the Different carries it (much more slowly) in the inverse direction along the ecliptic, *EPBQ.* So at the end of the twenty-four hours the center of the sun will not have got back quite as far as *Q*; it would only have reached a point *Q′*, a little to the east of *Q* on the arc *QE*. Hence the sun's real motion during that interval will not have been a circle on our sphere, but a very thin spiral on it, starting at *Q* and winding up at *Q′*. For the same reason in the next twenty-four hours it will have slipped back to *Q″*, a little further east along the arc *QE*, and another coil will have been added to the spiral which

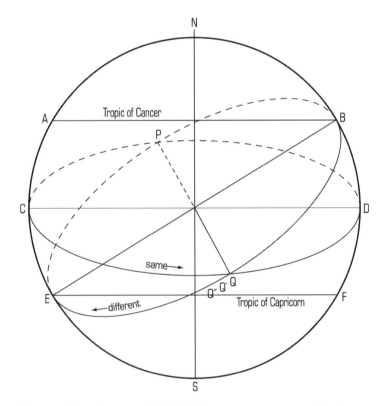

Figure 1. The distances QQ′, Q′Q″, have been exaggerated; had they been kept in scale, points Q, Q′, Q″ would have been too close to show up. (Adapted from Heath, *Aristarchus of Samos*, p. 160)

we started at *Q*. This will continue day by day, the spirals getting a little shorter at every twist, though always traversed in equal times, until the winter solstice is reached at *E*. Thereupon the process will reverse: the spirals will keep ascending, now getting successively larger until the vernal equinox, whereupon they will start getting shorter once again until the summer solstice is reached at *B*, and then the process will again reverse. We will thus get two sets of continuous spirals, one ascending from *E* to *B*, the other descending from *B* to *E*, one spiral for every calendar day, always described in equal times, but with varying lengths, smallest at the solstices, largest at the equinoxes.

So the value of Plato's hypothesis speaks for itself. A conceptual matrix which generates such information as I have just detailed leaves no doubt of its scientific value. And how did Plato happen to hit on it? So far as we can tell from the record, he got it via his metaphysical scheme: Since the sun is a god whose thoughts determine his movements, his soul must be perfectly rational (theses *A* and *C* above); and since rational motion is circular (thesis *D*), the sun's motion has to be circular (thesis *B*). Why then does not solar motion exhibit the simple, straightforward circularity of the movements of the fixed stars? Why is it not, like theirs, restricted to circular motion westward in a single plane? Why does it slip back eastward day by day? And why does the slippage occur in a plane which cuts obliquely across the planes in which the fixed stars move? Is not his soul as rational as theirs? Why then should his motion be so different from theirs? Plato's hypothesis answers successfully all these questions: hypothesize two movements in intersecting planes in different directions at different speeds, each of which operating singly would carry the sun in a perfectly circular orbit, and you can deduce that their concurrent operation will produce a helix which accounts for the phenomena[89] consistently with the demand of thesis *B*.

---

89. Plato must have disregarded the (pseudo-)phenomenon which caused Eudoxus to postulate a third sphere in his theory of solar motion, i.e., the belief, held "by Eudoxus and those before him" (the latter phrase would

Having gone that far, Plato could hardly go no further. He
would want to satisfy the same demand in the case of the
motions of the moon and the five planets. But here his hypothesis does not begin to give him what he wants. Here he
confronts motions which, unlike the sun's, cannot be accounted for by decomposing each of them into the movements of the Same and of the Different. The motion of the
Different in any member of the sextet should proceed either
in the ecliptic, as it does in the case of the sun, or else in a
fixed plane parallel to the ecliptic. Why then those variations
in latitude, those movements alternately approaching the
ecliptic and departing from it? These lateral motions are left
totally unexplained by Plato's hypothesis of the Same and the
Different. So are the variations in angular speed relative to
the sun exhibited by Venus and Mercury which Plato recognizes in describing the trio as alternately "overtaking, and being overtaken by, one another" (38D4–6) and, more generally, the retrogradations exhibited from time to time by each
of the five planets. From the standpoint of Plato's hypothesis,
which provides only for movements constant in speed and
direction, those intermittencies of planetary motion are scan-
dalous irregularities; they are phenomena which Plato's hypothesis cannot touch.

Plato's metaphysics offered him a facile way of coming to
terms with these phenomena. He could have said: 'Well, aren't
the moon and the five planets gods, with a mind and a will of
their own? Why then should they have to stick to those simple
patterns of motion which result from the composition of the
Same and the Different? Why shouldn't one or more of the
"wandering" stars choose to vary that one movement which is
observed so monotonously by the sun?' According to Cornford (*Plato's Cosmology*, pp. 8off., 1o6ff.), this is in fact how
Plato's astronomical theory does account for those phenomena of planetary motion which are recalcitrant to the

---

naturally cover Plato), that "the sun does not always rise in [exactly] the same
spot in summer or in winter solstices" (Simplicius, *in de Caelo* 493.16–17),
which would entail that the sun's orbit is at a slight inclination to the ecliptic
instead of coinciding with it.

hypothesis of the Same and the Different: Cornford would have us believe that the power of self-motion of the moon and of the five planets constitutes for Plato a "third force" (p. 87) which modifies in their case the direction and velocity imparted to each of them by the Same and the Different.[90] This proposal fails for want of textual evidence:[91] there is no suggestion in Plato's text that he ever had recourse to such a "third force" to explain phenomena unexplained by the Same and the Different.[92] And there is a very good reason why he should not—a reason which does not seem to have occurred to Cornford:[93] This is that the explanatory value of such an ancillary hypothesis would have been bogus. It would purport to explain observed irregularities in the motions of this or that star by postulating that the star simply chose to move in just those ways for just those periods of time. *Why* it should have made those choices rather than any of the infinitely many alternative ones it could have taken at those same points in its trajectory would remain a mystery. The proposed *explanans*, itself no less obscure than the *explanandum*, would explain nothing. So it is very much to Plato's credit that, so far as we can tell from his text—the only relevant source of information—he denied himself that all-too-easy way of resolving the huge discrepancies between the phenomena of lunar and planetary motion and his astronomical hypothesis of the Same and the Different.

An ancient tradition represents Plato as having put up to others the problem which he himself had failed to solve. It comes to us through this quotation from Sosigenes[94] in Sim-

90. For a convenient summary of the theory of planetary motion which Cornford ascribes to Plato in the *Timaeus* see his "Table of Celestial Motions," *Plato's Cosmology*, pp. 136–37.

91. It is rejected on this ground by Dicks, *Early Greek Astronomy*; see his extended critique of Cornford's theory, pp. 124–27.

92. I argue for this claim and raise a further objection to Cornford's proposal in Appendix, section K.

93. Or to any of his critics, to my knowledge.

94. The peripatetic philosopher of the second century A.D. to whom we owe a number of valuable comments (preserved in Simplicius' *in de Caelo*: see Diels's *Index Nominum s.v.* Σωσιγένης) on astronomical theories of the fourth century B.C.

plicius (*in de Caelo* 488. 21–24 [included in Eudemus fr. 148 Wehrli]):

> Plato had set this problem to those who were engaged in these [sc. astronomical] studies: What uniform and orderly motions must be hypothesized to save the phenomenal motions of the wandering stars?[95]

The historical truth of this report, which we have no way of nailing down,[96] is of itself of no great consequence. For regardless of whether or not Plato had ever formally proposed that task in so many words—these or others[97]—to the astronomers of his day, we can be certain that the task would pose itself to anyone who had come to know his astronomical theory and had the power to understand both its considerable achievement and its residual failure. Such a man, surely, was Eudoxus. Historians of astronomy have long recognized[98] that his theory of homocentric spheres[99] represents, if not a solution to the problem formulated in the above quotation, at least a brilliant advance in the right direction. What has not been recognized, to my knowledge, is the continuity between that advance—one of the greatest ever made in the history of astronomy—and the one represented by Plato's solution of a piece of the same problem in his theory of solar motion. In both cases we have composition of simple, uniform, circular motions accounting for complex, irregular, phenomenal motion. Applying this formula to the motion of the sun, Plato gets excellent results; applying it to the motion of the planets he gets poor results, for in their case his hypothesis is much too simple to cope with the complexities of the phenomena. Eudoxus applies the same formula to the motion of the planets and gets vastly better results, be-

95. Πλάτωνος . . . πρόβλημα τοῦτο ποιησαμένου τοῖς περὶ ταῦτα ἐσπουδακόσι, τίνων ὑποτεθεισῶν ὁμαλῶν καὶ τεταγμένων κινήσεων διασωθῇ τὰ περὶ τὰς κινήσεις τῶν πλανωμένων φαινόμενα.
96. For discussion of the relevant evidence see Appendix, section L.
97. For the phrase "saving the phenomena" see Appendix, section M.
98. See, e.g., Duhem, *Sōzein ta phainomena*, pp. 3ff.
99. For an exceptionally lucid elementary account see T. S. Kuhn, *The Copernican Revolution*, pp. 55ff., "The Theory of Homocentric Spheres."

cause he has transformed the simple Platonic construct, replacing it by a highly imaginative new model, where trios or quartets of homocentric spheres do the work of Plato's pairs of circles, bringing to the exploitation of this more powerful apparatus the resources of his mathematical genius. One could not, of course, compare the magnitude of Plato's scientific achievement with that of Eudoxus. It is nonetheless true that it was Plato who prepared the way for Eudoxus' triumph, for Plato's work had both posed the problem, and had already solved it in the much simpler case of the motion of the sun. In that case Plato had indeed shown "what regular and orderly motions must be hypothesized to save the phenomenal motions": the movement of the Same in the plane of the equator completed in just twenty-four hours and the movement of the Different in the plane of the ecliptic completed in just over 365 days.

I began this chapter by dramatizing Plato's hostility to the discoverers of the cosmos. His opposition to them was so savage that if they had turned up in his Utopia he would have silenced them by imprisonment or death. But what is the opposition about? It is not about the cosmos; it is about their theory of the cosmos—that is what he is determined to smash, though only to replace it by a new one of his own which would have been fully as obnoxious to the pietistic fanatics who prosecuted Anaxagoras and Socrates. For while Plato's cosmology makes fulsome acknowledgment of supernatural power in the universe, it does so with a built-in guarantee that such power will never be exercised to disturb the regularities of nature.[100] As I explained at the start, the guarantee is given tacitly, in the character of the supernatural creator of nature. He is a craftsman whose only wish is to make a world that will be as beautiful as his craft can make it. How then could his unenvying nature subsequently disrupt the order he put into the world to make it the beautiful thing it is? If

---

100. The same sanctions for impiety are to be applied to anyone who, believing that gods exist and care for men, thinks that "they can be bribed by prayers and sacrifices" (*Laws* 885B).

you accept the Platonic model of the cosmos you are no more
tempted to read an eclipse as a portent[101] or to interpret a
man's behavior as the result of *atē* than if you accept the
Democritean model. In either case you are assured of the
unfailing operation of what has since come to be called
"natural law." Only the ground of that law would be different: natural in the Democritean model, supernatural in the
Platonic.

So Plato's rejection of Ionian *physiologia* is not the total disaster it would seem to be at first sight. Even so, is it not a
mistake? If we had to choose between those two models of
the cosmos, would we not have every reason to opt for the
Democritean? Good reason—yes; every reason—no. Its neces-
sitarianism would have saved us from those preposterous inferences, legitimized by Platonic teleology, which decide what
is, or is not, the case, by deduction from what would be good
and beautiful, enchanting or inspiring, if it were the case. To
the extent to which the Platonic model incorporated such inferences it was wrong and, in the long run, was bound to be
junked by the scientific community. What is sometimes mis-
leadingly described by historians of science as "the dissolution
of the Cosmos" in the sixteenth century of our era is in fact
the dissolution of the *Platonic* model of the cosmos—the model
of a finite universe, bifurcated into two diverse realms, celestial
and terrestrial, governed by different types of motion, soul-
initiated circular motions for the heavens, mechanically caused
motions of other sorts for the sublunary sphere. This is the
world-picture invented by Plato and consolidated by Aristotle
that had to be scrapped before modern science could get under way.[102] So in the long run the choice for the Democri-

101. See the remark on 40C9–D2 (quoted in the text above, p. 50) about
celestial occultations which "bring terrors and portents of things to come *to
those who cannot reason.*"

102. "The dissolution of the Cosmos," writes A. Koyré (*Metaphysics and
Measurement*, p. 20), "means the destruction of the idea of a hierarchically-
ordered finite world-structure, of the idea of a qualitative and ontologically
differentiated world, and its replacement by that of an open, indefinite and
even infinite universe, united and governed by the same universal laws; a
universe in which, in contradiction to the traditional conception with its

tean, against the Platonic, model would be proved right—but in the very long run. In the short run the Platonic had distinct advantages, which, contrary to all expectations, made it in major respects more useful than its rival for the science of the day. I reached that conclusion reluctantly, almost incredulously.[103] The notions that the world has a soul, that all the stars have souls, and that celestial bodies move as they do because their rational souls must move in perfect circles are so fantastic that I found it hard to believe that they could have suggested a scientifically valuable systematization of scientifically established data. If the foregoing argument is correct, this is in fact what happened. The scientific theory of celestial motions suggested to Plato by his metaphysical scheme conveyed genuine insight, and had in any case distinctly greater scientific value than did the Democritean alternative for the same data. Seeking a mechanical explanation of the motion of the heavenly bodies, Democritus had turned to

---

distinction and opposition of the two worlds of Heaven and of Earth, all things are on the same level of Being." The "Cosmos" here described is precisely the *Platonic* cosmos, a fact which Koyré seems to ignore, representing it as an exclusively Aristotelian creation and categorizing its destroyer, Galileo, as a Platonist. Certainly there are Platonic features in Galileo's outlook. But in his protest against a "hierarchically-ordered," "closed," bifurcated universe, Galileo (whether he realized it or not) was fighting Plato fully as much as Aristotle.

103. These words were written in the summer of 1972, when the Danz lectures were delivered. It was nearly two years later that I read the section entitled "Die Entwicklung der antiken Astronomie" in von Fritz's *Grundprobleme der Geschichte der antiken Wissenschaft* (which was published in 1971 but did not reach me till much later; all of my references to this work were added in the final revision of my manuscript, during June 1974) and discovered to my joy that this great scholar, whose knowledge of Greek science is so much more profound than mine, had reached virtually the same conclusion: Having quoted (p. 180) the remark of a historian of mathematics "that Eudoxus shared Plato's most trivial 'philosophical' view that spheres are 'divinely' or 'transcendentally' beautiful," von Fritz observes: "Es ist aber gerade die von den 'trivialen' Vorstellungen der Göttlichkeit und Harmonie des Kosmos ausgehende rigorose Forderung der Gleichförmigkeit reiner Kreisbewegungen, welche die stürmische Entwicklung der griechischen Astronomie auf das kopernikanische System hin in Gang gesetzt hat. So paradox vom Standpunkt eines orthodoxen modernen Wissenschaftsbegriffes aus ist die antiken Entwicklung bei den Griechen verlaufen" (p. 182).

that old standby of the *physiologoi*, the vortex: the stars moved as they did because they remained in the grip of the vortical motion which, in the distant past, had caused the formation of our world; and the differences in their angular velocities were due to the fact that the vortex got "weaker" as it approached the center, so that the sun got "left behind" the fixed stars, and the moon still more so.[104] But the explanatory value of this hypothesis was illusory: there was no way of deriving from it testable correlations of differences in velocity between, say, moon, sun, and fixed stars with differences in their respective distances from the earth. The hypothesis was powerless to save the phenomena and to serve as a useful guide to new observation that would confirm the hypothesis or show the way to amend it.[105] Indeed it is hard to see how any dynamical theory that could be constructed with the technical resources of the time would have had good chances of success. Such a thing remained beyond the powers of the greatest astronomers of the West for two millennia. For this reason too Plato's

104. The Democritean theory, as paraphrased by Lucretius (*de Rerum Natura*, bk. 5, vv. 622-34): ". . . the nearer the several stars are to the earth, by so much less can they be carried along by the celestial whirl; for the rapidity and sharpness of its force diminishes lower down, and so, little by little, the sun is left behind the rearward constellations. . . . And the moon all the more so: the lower her orbit, the nearer to the earth, the less can she keep pace with the constellations. . . ." The vogue this theory enjoyed in Plato's time may account for the vehemence of his denial of its initial premise, i.e., that the fixed stars are the fastest celestial bodies, the moon the slowest; Plato asserts repeatedly that the very opposite is the case: it is not the fixed stars which "overtake" the moon, but the moon which "overtakes" them (*Ti.* 39A; *Laws* 822A–C); cf. n. 74 above.

105. The same criticism would hold with even greater force of the supplementary hypothesis (reported by Lucretius in the immediate sequel, vv. 627-43), to account for the concurrent seasonal movement of the sun from solstice to solstice: winds from the north hit it when it reaches the northern tropic to send it reeling back toward the south where it runs, just six months later, into a blast from the south which forces it to reverse. (Similar explanation of other movements of the moon and the planets [excursions in latitude and retrogradations?] in vv. 644–49.) If Democritus held this theory (we know that Anaxagoras did so along with the vortical explanation of the diurnal motions: Hippolytus, 1.6.8 and 9 [in DK 59A42]) its value for working astronomers would have been nil or worse. If they gave it any serious attention, which is doubtful, it would have been a sheer waste of their time. Cf. van der Waerden's criticism of the Anaxagorean theory, *Science Awakening*, p. 128.

metaphysical scheme would be an asset for the working astronomer: it would relieve him of the necessity of construct-
ing a dynamical model of the motions of the heavenly bodies.
It would license him to rest content with a purely kinematical
model which aimed to show how, if certain motions were assumed, the mathematically deduced consequences would
save the phenomena.[106] This was to be the way of the future—
the road traveled by Eudoxus, Apollonius, Hipparchus, Ptolemy.[107] What emboldened Plato to start that tradition was a
metaphysical fairy tale.[108] This was what saved him—and
through him the greatest astronomers of antiquity—from the
futile search for the physical causes of celestial motions by
denying that there were any such causes to look for, assuring
him that the observed facts were sufficiently accounted for by
the psychokinetic motions of the Same and the Different.[109]

106. I am at a loss to account for the introduction of "forces" in the eluci-
dation of Eudoxus' system of homocentric spheres by F. Laserre, *Die Fragmente des Eudoxus von Knidos*, pp. 163–64. I find nothing in the evidence to
support the suggestion that Eudoxus "had foreseen that the movement of
each planet along the orbit imposed on it by the layout of the spheres is
governed by two invariable and opposing forces."

107. This is how Ptolemy celebrates that journey's end: "Now that we are
about to demonstrate in the case of the five planets, as in the case of the sun
and the moon, that all of their phenomenal irregularities result from regular
and circular motions—for such befit the nature of divine beings, while disor-
der and anomaly are alien to their nature—it is proper that we should regard
this achievement as a great feat and as the fulfillment of the philosophically
grounded mathematical theory [of the heavens]" (*Syntaxis Mathematica* 9.2
[*Opera Omnia*, J. L. Heiberg, ed., vol. 1, part 2, p. 208]).

108. In all of my reading in the history of science I cannot recall a more
striking counterexample to what Karl Popper ("Back to the Presocratics," in
Conjectures and Refutations, p. 137) has called "the Baconian myth."

109. I acknowledge with deepest gratitude exceedingly useful criticisms of
parts of this chapter from my colleagues, Professors Thomas S. Kuhn and
Michael S. Mahoney, which have enabled me to correct a number of mis-
takes.

3

Plato's Cosmos, II:
Theory of the Structure of Matter

LET me reiterate one of the main points I made in the previous chapter: declared enemy and would-be persecutor of the *physiologoi* though Plato certainly was, he nonetheless accepted their discovery of the cosmos. In this crucial respect his world is like theirs and unlike that of Pindar, Herodotus, and of the vast majority of his own contemporaries: it is a world-system whose order is safe from supernatural interference. This is true even in his theory of the heavens, where his frontal attack on the *physiologoi* is deployed. Expunging physical causes from this region of the universe, he does not return it to its traditional Olympian masters. He installs another set of causes, psychic ones, to be sure, but designed to guarantee that the motions of the heavenly bodies will be as regular as any vortex or other mechanical agency could have made them. Plato's god-stars can no more "overstep their measures" than could the soulless fire-stars and earth-stars of the *physiologoi*.

When we move from the first division of the *Timaeus* (29E–47E), which deals mainly with the teleologically ordered motions of souls, to its second division (47E–69B), which deals with the mechanistically ordered motions of earth, water, air, and fire, the community of outlook which Plato shares with his adversaries becomes far more massive. The theory of the structure of matter which he presents here has so much in common with its counterparts in Ionian *physiologia* that Aristotle takes it for granted that Plato's theory is a variant of the atomic hypothesis of Leucippus and Democritus.[1] Plato himself invites comparison with the atomists, with Anaxagoras, and with Empedocles. To Anaxagoras he alludes briefly by speaking of the invariants of his own physical system as "seeds,"[2] which Anaxagoras had used as a technical term for that purpose.[3] His allusion to the atomists is more pointed and more sustained. It was they who had called the primary constituents of matter "*stoicheia*," "letters," analogizing their atoms to the characters from whose combinations and permutations the whole infinite corpus of sentences in an alphabetic script could be produced.[4] Plato borrows the term repeatedly: He calls *stoicheia* the elementary figures from which the corpuscles of his system are constructed.[5] On one occasion he complains that those who had posited earth,

1. See, e.g., how he juxtaposes the Leucippean-Democritean and Platonic theories of matter: both uphold "indivisible magnitudes" (i.e., maintain that the ultimate constituents of matter are atomic); but for the Ionian atomists these magnitudes are "bodies" or "solids," while they are "planes" for Plato (*de Generatione et Corruptione* 315B28ff., 325B15ff.).

2. When he assigns the tetrahedron, the octahedron, and the icosahedron, respectively, to fire, air, and water, he speaks of the first (and by implication of the other two as well) as "the element (στοιχεῖον) and seed (σπέρμα)" of its natural kind (56B). The use of "seed," after "element," for the primitives of the physical system, would be a stylistic redundancy if Plato were not alluding thereby to another famous use of this word for the same purpose.

3. Anaxagoras B4. For my interpretation of the significance of the term in his theory, see "The Physical Theory of Anaxagoras," pp. 31ff.

4. See Aristotle, *Metaphysics* 985B13–19, where the use of the analogy is spelled out: Leucippus and Democritus account for all the properties of matter by differences in the shape (ῥυσμός), concatenation (διαθιγή), and position (τροπή) of its constituent atoms; differences of shape are illustrated by that of the letters *A* and *N,* and of position by that of Zeta and Eta (in the earlier Greek script these were the same character rotated 90°). Though this is, of course, Aristotelian paraphrase, not quotation, there can be little doubt that it reproduces authentic atomistic vocabulary and imagery; the Democritean use of the word ῥυσμός for atomic shapes is confirmed by the title of one of his works, περὶ τῶν διαφερόντων ῥυσμῶν (Democritus B5i): it is commonly agreed that this dealt with differences in atomic shape. For interesting comment on these terms see K. von Fritz, *Philosophie und Sprachlicher Ausdruck bei Demokrit, Plato, und Aristoteles,* pp. 24ff. Von Fritz points out that Democritus' earlier contemporary, Herodotus, had used ῥυθμός with reference to the characters of an alphabetic script, remarking that the Greeks altered τὸν ῥυθμὸν τῶν γραμμάτων when they adopted an alphabet from the Phoenicians.

5. 54D6, 55A8. See also the citation from 56B in n. 2 above.

water, air, and fire as the *stoicheia* of the universe had not
penetrated very deeply: what they had taken as the letters of
the alphabet of nature should not be ranked, he says, even
as low as syllables for this purpose.[6] Here he must have had
Empedocles in view,[7] for it was he who had made earth, water, air, and fire the elements of the natural universe, endow-
ing them with absolute, Parmenidean, unalterability to make
them qualify for this purpose. So in offering us his own *stoicheia* Plato shows that he has enrolled in the same program
of physical elementarism. Like Empedocles, Anaxagoras, and
the atomists, he is undertaking to identify those recurrent
elements whose shufflings and reshufflings would explain the
law-like constancies of the flux. He wants us to know that
the job his predecessors had tried to do has now been done,
properly, by himself: the letters, not just the syllables, in
which nature writes her narrative have been made out at last.

But in all this we have yet to put our finger on Plato's special affinity to the atomists—or, to put it in less flattering
terms, his unacknowledged debt to them, which is so much
greater than what he owes to any other natural philosophers.
To get at this we must remind ourselves of what it is about
the atomists which constitutes the greatness of their guess at
the riddle of matter. This is surely not just the fact that they
posited indivisibles. It is the deeper insight they revealed in
making this posit, namely, that the unobservables we postulate to account for properties of observables need not themselves possess those same properties. Divisibility is a case in
point. We do impute it, and with good reason, to every material object that falls within the range of our senses. Must we
then also impute it to the invisible, intangible, micro-entities
which make up the things we see and touch? Why should we?
So far as logic goes, there is no warrant for reasoning, 'X is
made up of a million Y's; X is divisible; *ergo* each of the Y's

6. 48B–C: "For no one has yet explained their origin but we say that they
are the primary entities (ἀρχαί), as if we knew what fire is and what each of
the rest is, while for one who had even the least intelligence they would not
deserve to be ranked even as low as syllables."

7. Though without implying that Empedocles himself had applied the term
"*stoicheion*" to the elements (which he had called "roots").

is divisible.' Recognizing the fallacy in that inference—"the fallacy of division" the logic books call it—we are left free to investigate the conclusion on its own merits and may reject it with an easy conscience if we find, as the atomists were convinced they did, reasons against it. By the same token, why should we assume that if *X* is red, each of the *Y*'s that make it up must be red, and if not red, then of some other color? And so the atomists find their way to *stoicheia* which are fantastically different from the familiar objects of sense-experience—elements destitute of all sensible properties except extension and hardness, and discharging all the better on that account their explanatory functions. This liberation of the theoretical imagination was the finest legacy of the Greek atomists to modern science. Plato must have sensed its value, for he went further, much further, along the same road. The indivisibles of his physics were still more remote from the bodies of sense-experience: they were not even bodies, but only bounding surfaces of bodies. His atoms are two-dimensional.

For a full account of his theory I can refer to F. M. Cornford's masterly discussion of it in *Plato's Cosmology*.[8] All I need do here is to run over the bare essentials, sketching swiftly the design that must be kept in view in the critical discussion of the theory which is to follow.

The matter which confronts the Demiurge in its primordial state is inchoate.[9] The four primary kinds of matter, earth, water, air, and fire, are present here in a blurred, indefinite, form; their motion is disorderly. The Demiurge changes all

8. But readers should be warned that parts of Cornford's reconstruction of the physical theory of the *Timaeus* are highly conjectural (see n. 17 below). For a recent critique see W. Pohle, "The Mathematical Foundation of Plato's Atomic Physics," pp. 36ff. Being as yet unconvinced that any proposed alternative is preferable to Cornford's reconstruction, I am adhering to its main lines in this paper, hoping to address myself before too long to some of its riskier features.

9. A majority of modern commentators have taken the view that this reference to a pre-existent chaos and the whole of the creation story in the *Timaeus* is not meant to be taken literally. I have defended the position I am taking here in two papers, "The Disorderly Motion in the *Timaeus*" and "Creation in the *Timaeus*: Is It a Fiction?" (in R. E. Allen, ed., *Studies in Plato's Metaphysics*, pp. 379ff. and 401ff.).

this. He transforms matter from chaos to cosmos by impressing on it regular stereometric form.[10] When he has done his job all of the fire, air, and water in existence will consist of tetrahedra, octahedra, and icosahedra respectively, that is to say of solids whose faces are invariably equilateral triangles. And earth will be found to consist of minute cubes. So far the difference from Democritus is not too great. Had Plato stopped here, all he would have given us would be a refurbished corpuscularism—the Democritean atom, deprived of its infinite multiformity, restricted to four shapes corresponding to four of the five regular convex solids of Euclidean geometry, to whose formal demonstration Plato's young friend, Theaetetus, had made important contributions. But that is scarcely the story Plato would be content to tell. His aim is not to improve the Democritean atom, but to junk it. He wants to open the way for the very thing which the unalterable solidity of the materialistic atom had been designed to block: the intertransformability of elementary corpuscles. On the Democritean theory an atom of fire, for example, could never change its shape or size in any way whatever—hence never change into an atom of air or of water. And this is precisely the sort of change Plato wants to insure. He wants corpuscles which will be susceptible of two types of radical transformation.

The first involves changes from one kind of matter into another—from fire into air or into water or both, and conversely. Since all three of these have identical faces—equilateral triangles which I shall call "t's" for

---

10. "Fire, water, earth, and air possessed indeed some vestiges of their own nature, but were altogether in such a condition as we should expect for anything when deity is absent from it. Such being their condition, he [the Demiurge] began by giving them a distinct configuration by means of shape and numbers" (53B1–5; translation adapted from Cornford). It is no use asking how the Demiurge manages to carry out this stupendous operation. We are dealing here with a strictly *supernatural* event (the event which creates nature and is not itself a member of the sequence of events that constitute the natural order). To offer any account, no matter how conjectural, of the reduction of the material chaos into a beautifully structured cosmos by an extramundane Intelligence would be as futile as an attempt to figure out the means by which a kind fairy transforms a pumpkin into a coach-and-four.

short—this type of intertransformation would have been assured if he had made t's the atoms of his theory, along with squares ("s's" for short) which form the differently shaped faces of the earth-cubes. For then all allowable transformations—that is to say, between tetrahedra, octahedra, and icosahedra, never between any of these and cubes—could be represented as the dismantling of n-faced polyhedra and reassembling them as m-faced ones. Here are some examples of atomic intertransformations permitted by the theory, with conjectural physical interpretations,

when W is used for a water-corpuscle (icosahedron),
A for an air-corpuscle (octahedron),
F for a fire-corpuscle (tetrahedron),
and t for the equilateral triangle which constitutes
each face of the above polyhedra, then

(a) $1W \longleftrightarrow 5F$
        $20t = (5 \times 4t)$

Physical interpretation (reading the formula from left to right): a burning oil-lamp. The icosahedra of the water component of the oil (which is taken to consist of water, with a little fire mixed in)[11] are being dismantled and reassembled as tetrahedra. Moreover,

(b) $1W \longleftrightarrow (3F + 1A)$
        $20t = [(3 \times 4t) + 8t]$

Physical interpretation (reading the formula in the same way): rapidly boiling water, turning into extremely hot air. Each icosahedron is being reassembled as three tetrahedra and one octahedron to constitute a mixture which is very hot, since it contains three corpuscles of fire for each corpuscle of air. Again,

(c) $1W \longleftrightarrow (1F + 2A)$
        $20t = [4t + (2 \times 8t)]$

Physical interpretation (reading the formula in the same

11. Liquids are plausibly construed as varieties of water by Plato in the *Timaeus* (60A–B), as generally by the Greeks.

way): slow evaporation of water warmed by the sun. Each icosahedron is changing into one tetrahedron and two octahedra to produce a mixture containing one corpuscle of fire for
every pair of air corpuscles—warm, but not hot. Finally,

(d) $2W \longleftrightarrow 5A$
$(2 \times 20t) = (5 \times 8t)$

Physical interpretation (reading the formula from right to
left): cold air turning into equally cold water.[12] Each quintet
of octahedra is changing into a pair of icosahedra.[13]

The second type of transformation Plato wants his theory
to accommodate is between varieties of each of the four primary kinds of matter. These would all be explained as "isotopes" of air, if I may follow Paul Friedländer[14] in lifting this
term from modern chemistry to denote something similar in
Plato's physical theory: corpuscles with the same shape, hence
with broadly similar properties, coming in different sizes,
each size meant to account for the special properties of particular varieties. Thus in the case of air Plato names two varieties—one of them popularly called "aether," the other "mist
and darkness" (ὁμίχλη τε καὶ σκότος)[15]—and says that there
are also other "nameless kinds" (58D); each would consist
of octahedra of different sizes: the smallest would constitute
the bright, clear, uppermost part of the atmosphere, the

12. Since there is no fire in either—only octahedra in the air, only icosahedra in the water.

13. For further implications of the theory see Appendix, section N.

14. *Plato,* vol. 1, *An Introduction,* p. 255. He emphasizes the liberties he is
taking in transferring the modern term to the radically different Platonic
context:

> I am not sure whether I am obscuring or clarifying the problem by giving the mod-
> ern scientific name "isotopes" to those different kinds of the same element [in the
> Platonic theory]. For many elements are thought of today as having different iso-
> topes: the isotopes of one element "are identical as to both the number and the
> arrangement of the electrons" in each atom; they "are quite indistinguishable from
> each other except for those properties which depend only on mass" [source of quo-
> tation not given]. Analogously, Plato's different species, say, of water or, better, of
> liquid (e.g. wine, oil, honey, acid) differ only in the size of their corpuscles, not in
> their form, which is always icosahedral.

15. For my claim that Plato means that these two isotopes of air are *popularly* so called see Appendix, section O.

largest one dark vapor.[16] To enable transformations between such varieties Plato constructs the t's—the equilateral triangles which bound the octahedra—from smaller triangles (see Figs. 2–5).[17] Here at last we get down to the rock-bottom, atomic, elements of Plato's theory. Every t is made up of scalene right-angled triangles whose three sides are in $1:\sqrt{3}:2$ ratio. Call these triangles "a's" for short. Two p's put together to form an equilateral triangle will make up the smallest t; six and eight a's will make up respectively the next two larger grades if put together in the ways indicated in Figure 5; still larger ones may be constructed by putting together eighteen a's, twenty-four a's, and so forth. But these cannot be the only atomic figures in the theory, for the square faces of the earth-cubes cannot be constructed out of these. So Plato postulates another element, the isosceles right-angled triangle whose sides are in $1:1:\sqrt{2}$ ratio. Figures 6 and 7 show how these may be put together to form faces of the first grades of earth; larger grades may be constructed by putting more of them together to form larger squares.

This is the gist of the mathematical, combinatorial, part of the theory. It is supplemented by another section on the mechanics of the transformations. Here I must abbreviate still more brutally, contenting myself with pointing out that Plato

16. That the first variety must be the smaller, the second the larger, I get by taking the statement that they are "brightest" (εὐαγέστατος) and "darkest" (θολερώτατος), respectively (58D), in conjunction with Plato's color theory (67C–68D), which correlates perceptible differences of bright and dark with the differences in corpuscular size, making the smaller corpuscles produce the brightest colors. I take Plato to be saying that of the many varieties of air the brightest and the darkest have given rise to the popular view that there are two stuffs, one of which is called "aether," the other "mist and darkness."

17. Figures 2, 3, and 4 have the advantage of showing how the faces of these three polyhedra—the t's—are constructed out of the elementary figures—the a's. Figure 5 shows separately the construction of t's out of a's. Following Cornford's interpretation (*Plato's Cosmology*, pp. 230ff.) the series of t's is depicted as starting with grade A (where $t = 2a$) and is continued for two more grades. It should be noted that the text (54D–E) mentions only t's made out of six a's, i.e., only the t's in grade B in Cornford's reconstruction. Grades A, C, and further members of the series are inferred, mainly from the meager data provided in 57C7–D5.

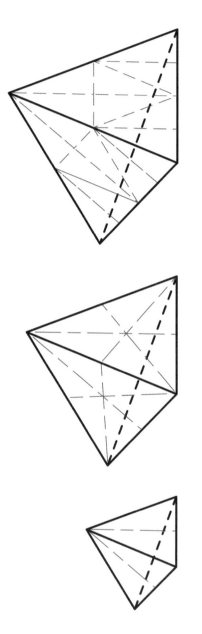

Figure 2. Regular tetrahedra: corpuscles of fire (taken from Friedländer, *Plato*, vol. 1, *An Introduction*)

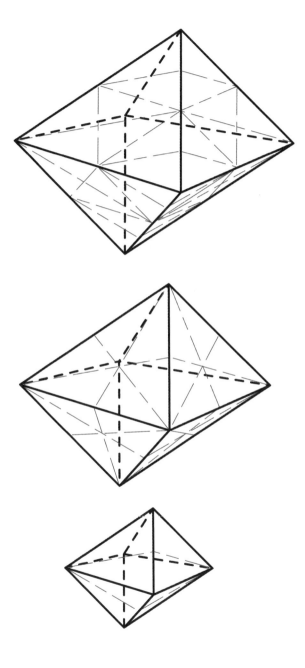

Figure 3. Regular octahedra: corpuscles of air (taken from Friedländer, *Plato*, vol. 1, *An Introduction*)

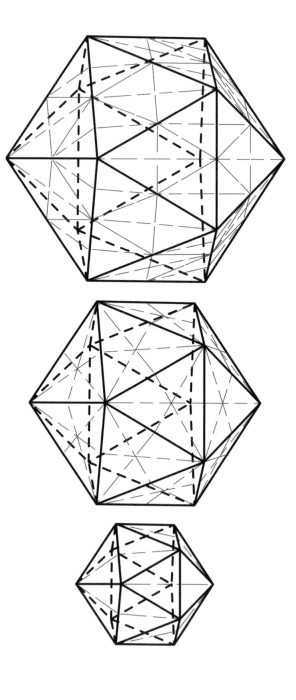

Figure 4. Regular icosahedra: corpuscles of water (taken from Friedländer, *Plato*, vol. 1, *An Introduction*)

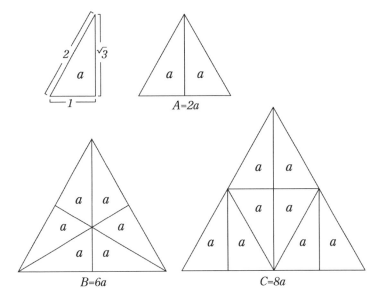

Figure 5. The construction of different grades of equilateral triangles from the elementary half-equilateral (a). The next term in the series would be $D = 18a$.

denies action at a distance[18] and reduces all physical interactions to operations involving contact—ultimately pushing, cutting, and crushing. Fire corpuscles cut air and water corpuscles, for the solid angles of tetrahedra are smaller—hence sharper—than those of octahedra or icosahedra; for similar reasons air corpuscles cut water corpuscles.[19] Here we see air changing into fire, and water changing into fire or into air or into both. The converse transformations will occur when a small quantity of fire is enveloped in a larger mass of air or of water, or a small quantity of air in a larger mass of water; then the bigger mass squeezes and smashes the polyhedra of the smaller one; we then see fire turning into air or into water or into both, or air turning into

18. No body moves another by "attraction" (ὁλκή): even magnetic phenomena are to be explained without making such an assumption (83B–C).
19. 56D–57A.

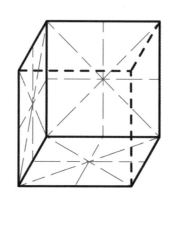

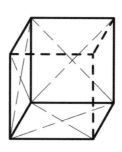

Figure 6. Regular cubes: corpuscles of earth (taken from Friedländer, *Plato*, vol. 1, *An Introduction*)

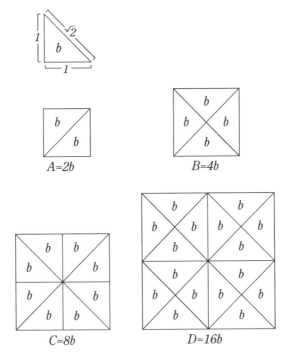

Figure 7. The construction of different grades of
squares from the elementary isosceles triangles (b)

water.[20] I skip further points in this section to raise at once
the question that specially concerns me: what is the value of
such a theory? In particular, does it have any scientific value?
Are we really entitled to call it a *scientific* theory of the constitution of matter?

Now the first thing we would like to know when assessing
a physical theory is: what are the facts which it purports to
explain? What are they in the present case? Plato alludes en
bloc to what he takes to be the explananda of his physical
theory. He does so at the very moment at which he starts his
stereometric construction in 53D–E. He asks there: "What are
the most beautiful bodies that can be constructed, four in

20. 57A–C.

number, unlike one another, but such that some of them can
be generated out of others by resolution?" What he means
by the latter phrase he had stated more carefully and more
fully in an earlier passage:

> First of all, take what we now call "water." When this is solidified
> we see, *as we think* (ὡς δοκοῦμεν), that it becomes earth and
> stones while, alternatively, that this very same thing [*sc.* water],
> when dissolved and dispersed, becomes wind and air, and then
> fire when the air burns up; and that fire, conversely, when con-
> densed and extinguished, goes over into the form of air; and that
> air again, when congesting and thickening up, becomes cloud
> and mist; and that from these, when they are further compacted,
> comes flowing water, and then from water come earth and stones
> once again;[21] thus, *as it seems* (ὡς φαίνεται), they pass on to one
> another in a cycle the process of generation. [49B7–C8][22]

What is the point of the two phrases I have italicized? Why
does Plato say, "we see, as we think . . ."? Again, in the
terminal sentence, what is the force of "as it seems . . ."? His
most immediate reason for speaking with such caution

21. I have divided into cola Plato's unbroken nine-line period, as has generally been done by his modern translators. The reader must nonetheless keep
track of the fact that all of the verbs denoting physical processes (participles
in Plato's Greek) are grammatically dependent on the initial phrase "we see,
as we think" (ὡς δοκοῦμεν, . . . ὁρῶμεν). I have taken care to indicate this in
my translation by putting all of those verbs into subordinate clauses. Some
modern translators (e.g., Bury and Cornford) have been as scrupulous, but
others have not (e.g., Archer-Hind, Apelt, Rivaud, Lee), thereby inadvertently
fostering the impression that Plato himself in this sentence may be subscrib-
ing to the physical reality of all of those changes—which is unfortunate, since
(as I proceed to explain in the text) his theory emphatically denies the change
from water to earth which is asserted in the penultimate colon in the same
modality as at the start: here as there Plato is not saying that from water
comes earth, or that we *see* this happening, but that *we think* we see it.

22. The "cycle of generation" to which Plato refers us here he takes over
from the *physiologoi.* It had figured prominently in their teachings as far back
as the sixth century. Thus, we find it in Anaximenes in the last third of the
sixth century in virtually identical terms: the downward phase of the cycle is
reported for him (Theophrastus, *ap.* Simplicius, *in Phys.* 24, 26) as fire, air,
cloud, water, earth, stones. In this cycle we get an ordering of the entities
which were generally taken to be the main stuffs that make up the matter in
the world: fire at one end, earth and stones at the other, air and water in
between. Further stuffs may be intercalated, the usual place for these being
the part of the cycle between water and air: we see Plato mention here two
forms of vapor, "cloud and mist," on the downward phase of the cycle.

should be apparent from the foregoing sketch of his theory of the constitution of matter: the theory will not allow all of the interchanges mentioned in the quotation. It allows some—those between water, air, and fire, whose surface boundaries are all composed of interchangeable atomic triangles. But it disallows others: it precludes interchanges between any of these three and earth, since the cubical corpuscles of the latter cannot be generated from the surfaces of tetrahedra, octahedra, or icosahedra. Anticipating this implication of his theory, Plato naturally holds back from saying that we *see* water becoming earth or earth becoming water: according to his theory there is no such thing to be seen. Aristotle is very critical of this part of Plato's theory:

> The first difficulty in [the Platonists' theory of the intertransformation of earth, water, air, fire by] the resolution of the plane-surfaces [of these bodies] is in [its] disallowing all [of those bodies] to be generated out of one another, which is what they are compelled to say, and what they do say. For it is not reasonable that one only [of these four bodies] should have no part in the transformations, *nor is this in accord with sensory appearance*[23] [which is] that all alike change into one another. Thus it turns out that, while discoursing about the phenomena, they make statements which are inconsistent with the phenomena. [*De Caelo* 360A1–7]

The charge is that when Plato confronts sensory evidence that falsifies his theory, he defies the evidence to stick by his theory. Is the charge fair? It would be if, and only if, it were the case that the intertransformation of fire, air, and water with earth were itself fixed in, or by, the sensory evidence. If that were the case, the consequences would be devastating not only for Plato but for most of the *physiologoi* as well: thus Empedocles, Anaxagoras, and the atomists had all believed that the primary constituents of fire, air, water, and earth were unalterable. On each of these theories if we were to take a primary unit of any of these bodies and put a tracer on it, we would find it absolutely unchanged for ever after: if it

---

23. οὔτε φαίνεται κατὰ τὴν αἴσθησιν. Aristotle's charge is brought out even better by the freer rendering in the Oxford translation of the *de Caelo* (by T. L. Stocks) of the phrase I italicized above: "and also contrary to the observed data of sense."

was, say, earth to begin with, it would never be anything but earth thereafter. Now if what Aristotle says were true—that "sensory appearance," that is, the sensory evidence itself, declares that earth, water, air, and fire "all alike change into one another," each of these theorists no less than Plato would stand convicted of holding a theory that clashes with the phenomena it is supposed to save. But Aristotle's allegation is in fact baseless. We do not know of any phenomena, observed or observable at the time, which would have falsified the theory of the unalterability of all the primary components. Those who maintained the theory would stand ready to show that any apparent facts to the contrary would fit their theory perfectly. To illustrate let me go back to the case of the burning oil-lamp, which could have been thought—and would be thought by Heraclitus, Plato, and who knows how many others—to involve the transformation of water into fire:

What are "the hard facts" in that case, if I may use this expression, here and hereafter, to mean facts ascertainable by observation on which each of the parties in a theoretical dispute could be expected to agree regardless of their other differences? Simply the following:

1. That the liquid in the container is that special stuff, commonly called "oil," identifiable as such by ordinary tests—those used for domestic, industrial, and commercial purposes;
2. That the volume of this liquid keeps dwindling as the flame in the wick burns on, and when the liquid is used up the flame sputters out and dies.

Now is there any difficulty in staying in line with these facts while maintaining with Empedocles, Anaxagoras, and Democritus that the transformation of water into fire is a physical impossibility? None whatever. In the absence of any surviving texts of theirs bearing directly on this question, we can still figure out how each might have explained the phenomenon consistently with their belief that water-atoms and fire-atoms are immutable from eternity to eternity. Take

Democritus, for example. The flame, he would tell us, is a column of fire-atoms. The ones it is constantly losing to the air around it are being replenished from fire-atoms which are scattered throughout the oil; when the flame is lit, these fire-atoms start streaming into the wick from all sides and keep it up so long as their supply holds out. And if we then ask why the whole of the liquid—not just its fiery content—is diminishing throughout this process, Democritus would have no trouble in spinning out an appropriate reply. For example, he might say that the water-atoms in the oil, agitated by the movement of the fire-atoms towards the wick, get shaken loose and rise to be dispersed in the air above in the form of a very thin, invisible, vapor, so that by the time all the fire-atoms have been used up, all the rest of the liquid in the container has cleared out as well.[24] So Democritus would be in a position to say that all the facts we do see are perfectly accounted for by his theory, and that the same is true in all the cases which Aristotle misdescribes as our seeing—or otherwise sensing—intertransformations of bodies like fire, air, water, and earth. In every case, Democritus would say, the sensible phenomena can be explained as the rearrangement and redistribution of absolutely unalterable micro-entities, that is, of things which cannot be themselves perceived but whose presence in what is perceived suffices to account for all perceptible mutations.

Similarly Plato would reject the assumption that the change of earth into water or air or fire is something we see. He could do so without having to deny a single hard fact[25] Aristotle could have produced to support that assumption. The liquefaction of metals in high temperatures would not qualify, for Plato would not reckon metals in their normal, solid, state as earths; he classifies them (58Dff.) as "fusible waters," that

24. This suggested explanation of the phenomenon by Democritus' theory is pursued further in Appendix, section P.

25. The reader may hark back at this point to my description of "scientifically ascertained facts" in Chapter 2, p. 36 above: these would be a subclass of what I am calling "hard facts" here—a critically important one, representing facts reached by scientific investigation.

is, as liquids with very high freezing-points. Nor would the fact that earth may be reduced to fluid form, as in volcanic lava, count, for Plato would say that matter in that state is earth in which "moisture had been left"[26]—that is, a mixture of earth and (fusible) water. The solubility of stuffs like salt and soda would not count either, for though Plato classifies these as varieties of earth,[27] he would say that, when dissolved, earth is not transformed into water but only dispersed in it, earth corpuscles being redistributed uniformly among much greater masses of water corpuscles. Nor finally would the combustion of solids count as earth turning into fire, for the only solids used as fuel in antiquity were wood and wood derivatives, and Plato would say that it is the water, the air, and the fire in wood and charcoal that burns; the earth content in these stuffs, he would say, is noncombustible ash. Plato then could claim, with reason on his side, that the question of the intertransformability of earth with each of the other kinds of matter is left open by the sensory evidence and should, therefore, be decided on theoretical grounds. He could so decide it without fear of empirical falsification, for there would be no hard facts to say him Nay.

I have lingered so long over Aristotle's objection, and the reply to it, because it dramatizes an issue of the greatest methodological interest. What is at stake here is the status of the primary data of physical theory in Plato and the *physiologoi.* All are in the business of producing theories about physical phenomena. So these phenomena must provide the empirical controls on theory-construction. To these each theorist would have to refer to vindicate his own hypothesis against those of his competitors, claiming that at this or that point the data yield confirmation for his and disconfirmation for theirs. If so, these phenomena cannot be simply what we "think" we see, or "seem" to see; they must be what I have been calling "hard facts"—statements on which sponsors of rival theories may expect to reach agreement by means of observations which are open to all and are not logically de-

26. See *Ti.* 60D3–4 and Cornford, *Plato's Cosmology*, p. 256, n. 2.
27. *Ti.* 60D4–E2.

pendent on the very theories under dispute. I am making provision here for two qualifications, apart from which asking for hard facts would be asking for the moon; I am allowing that the facts in question need not be derived by a process which either (1) has a built-in guarantee against the possibility of error, or (2) is itself innocent of theoretical commitments. The first is not required: all that is necessary is to be on guard against the hazard of error and to have the means of narrowing it down by repeating or varying the observations. Nor is the second necessary; all that is needed is innocence of commitment to the very theory which is under debate. This is fortunate, for if one had to do without theory altogether one's observational ventures would be doomed to triviality: the interesting observations are precisely those which constitute responses to questions raised by theory and are recorded in language provided by theory.[28]

In astronomy, to recall what I said in the second chapter, a corpus of satisfactory "hard facts" was being assembled: by the latter part of the fifth century people like Meton and Euctemon were doing the job. Who was doing the corresponding job in the area of terrestrial physics? Nobody. So the only data available to a physical theorist like Democritus or Plato would be those of ordinary, uncontrolled, unrefined, unanalyzed observation—things which everybody was supposed to know and no one was expected to investigate. These data were powerless to decide between rival theories such as those of Democritus and Plato. Compare the explanation of the workings of an oil lamp we could expect from Plato with the one I allowed Democritus a few pages above: the tetrahedra that form the base of the burning wick dissolve icosahedra in the oil in its immediate vicinity, thereby transforming them into tetrahedra; most of these move into the wick, while a few move in other directions, increasing the fire

28. To require the empirical scientist to meet conditions (1) and (2) above would be to cripple his inquiries. In the account of "scientifically ascertained facts" in Chapter 2 (p. 36) the second and third of the conditions given there show how arbitrary it would be to impose on scientific inquiry these crippling constraints.

component of the oil and warming it.[29] How could the facts
at the disposal of these rival theorists prove one of them right
against the other? To decide the case for Plato the facts would
have to show, for example, that the liquid has a water content
which consists of icosahedra, and that the fiery part of the
burning wick consists of tetrahedra. How could the observable facts to which he could appeal satisfy skeptics that his
imaginative tale has physical truth? Or how could they do
better for Democritus? The facts of observation, such as they
are, will fit both theories, hence cannot support either against
the other. So each of the two constructs is safe from refutation by experience and for that very reason incapable of confirmation from it.

This, I submit, is the basic predicament of physical theory
in the *physiologoi* and in Plato—they are all in the same boat
in that respect. This predicament has not been clearly under-

29. Let me run over the questions arising for Democritus (see Appendix,
section P) to see if any of them could cause Plato any embarrassment:

1. How is it that the temperature of the oil may vary while the ratio of fire
to water in it remains constant? For Plato this question does not arise: in his
theory the fire-water ratio does *not* remain constant under different temperatures.

2. Why do fire-atoms generated in the oil move massively into the burning
wick? Plato's answer would be that fire-atoms have a tendency to move toward "their own kindred" in the higher regions of the atmosphere (62C–
63E); hence terrestrial atoms have a general tendency to move upward;
hence the bulk of the tetrahedra generated next to the base of the burning
wick will rush into the upward-moving column of fire-atoms that form the
flame, thus replenishing the supply it is constantly losing to the air above
and around it.

3. If water-atoms are escaping upward why is there no visible vapor? For
Plato this question arises no more than does the first: in his theory all of the
heat above the oil is accounted for by tetrahedra that have been generated
from octahedra.

It should be hardly necessary to emphasize that the account of the Platonic
explanations I offer here and in the text above is heavily conjectural: Plato
has left us no description of this phenomenon, nor of any other which would
enable us to infer with reasonable certainty how he would handle this one. I
have had to base my account on such slender textual data as the description
of oil as a mixture composed of water and fire (60A), the reference to the
dissolving effect of fire on water (56E), and the implication that water loses
its fire-content when it congeals (56D–E).

stood in the scholarly literature.[30] Some of the most distinguished figures in the field have missed it completely. Thus F. M. Cornford, after life-long labors in this area, put his name to the following description of the *physiologoi:*

> It is not only that they describe, with complete confidence, matters beyond the reach of observation, such as the origin of the world; but when they come to matters of detail, they make assertions which could have been upset by a little careful observation or by the simplest experiment.[31]

If that is what you believe about the *physiologoi,* you would have to think them either simpletons, incapable of "a little careful observation," or else ostrich-headed dogmatists, unwilling to make the observations that would "upset" their theories. So far as we can tell from the evidence, neither of these things is true: we do not know of a single fact observable by the *physiologoi* with the means at their disposal which could "upset" any major theory of theirs in the domain of terrestrial physics. The ones adduced by Cornford are bogus: they would not even dent the theories they are supposed to crack, as I pointed out briefly some twenty years ago when I reviewed his last book.[32]

30. Or, if it has been understood, it has not been effectively communicated. Consider: "Empedocles did not allow the transformation of one root into another, and yet it is a matter of common experience that water, for example, becomes vapor (that is, in the Greek view, ἀήρ) on being heated to boiling point, and that this ἀήρ may recondense and become water again" (G. E. R. Lloyd, "Plato as a Natural Scientist," p. 85). The writer fails to explain that the nontransformability of air into water in Empedocles' theory comports perfectly with the "common experience" of water becoming vapor when sufficiently heated, and vapor becoming water again when it is cooled: vapor for Empedocles is simply air saturated by water particles.

31. "Was the Ionian Philosophy Scientific?" *Journal of Hellenic Studies,* pp. 1ff. This paper was written shortly before Cornford's death. It summarizes the argument which he presents in his posthumously published *Principium Sapientiae* (see next note).

32. Review of *Principium Sapientiae* (1952) in *Gnomon* 27 (1955): 65–76, reprinted in Furley and Allen, eds., *Studies in Presocratic Philosophy* 1:42ff. Thus Cornford had argued that the Empedoclean theory of respiration (that air goes in and out of our bodies through "pores") could have been refuted "by anyone who would sit in a bath up to his neck in water and observe whether any air bubbles passed into, or out of, his chest as he breathed" (*Principium*

As you go through the physical theories Plato writes into the *Timaeus*, empirical objections of all sorts keep popping in your mind. Time and again it looks as though he is saying something whose consequences are flagrantly, outrageously, out of line with observable facts, and you wonder why Plato does not seem to notice. But if you track down the question you will probably find that he would have anticipated no such trouble. Let me illustrate with a major example—one which concerns the compatibility of the fundamentals of his stereometry with certain empirical data which were well known at the time, as is clear from this remark in Aristotle's *de Caelo* (305B11-15):

> When air is formed from water, it takes up greater room, for the finer body [i.e., the one which is composed of finer particles] occupies more room. This becomes evident in the process of change: as the liquid turns to vapor and to air, the vessel in which they are enclosed bursts because it does not have enough room.[33]

What would this do to Plato's theory, where it follows by geometrical reasoning that when icosahedra turn into octahedra there is not gain, but loss, in aggregate volume, and even greater loss when icosahedra turn into tetrahedra, as specified in Table 1.

Consider again case (b) (see p. 71 above)—evaporation of rapidly boiling water—where each icosahedron is turning into three tetrahedra and one octahedron, thereby *losing* nearly two-thirds of its volume (same table). How reconcile

---

Sapientiae, p. 6). The force of the supposed refutation depends entirely on the assumption that air breathed out of our body into the bath water would clot into bubbles instead of being uniformly dispersed in the liquid. This bit of armchair science was so firmly lodged in Cornford's head that it never occurred to him that Empedocles might have not shared it. There is nothing whatever in Empedocles' known theories to commit him to it.

33. Aristotle brings up these facts not against Plato (he does not advert to their implications for the stereometric theory of the *Timaeus*) but against the physical theories of Empedocles and Democritus. That the facts to which he refers are matters of common, long-standing, knowledge is a reasonable inference from the way he introduces them: had he thought them new discoveries, unknown to fifth-century thinkers, he probably would have said or implied as much.

TABLE I

VOLUMES OF PLATO'S POLYHEDRA, STATED AS FUNCTIONS OF s
(Side of their equilateral triangular faces)

Volume of Icosahedron = $2.1817s^3$
Volume of Octahedron = $0.4714s^3$
Volume of Tetrahedron= $0.1178s^3$
Aggregate volume of $3F + 1A = 0.8248s^3$

Source: E. M. Bruins, "La Chimie du *Timée,*" *Révue de Metaphysique et Morale* 56 (1951): 269ff.

this with the fact—a good "hard fact" in this case—recorded by Aristotle in the above citation?

Plato could hardly have been unaware of either (a), the empirical data to which Aristotle alludes, nor yet of (b), the volumetric consequences in his stereometry of the conversion of water into a blend of fire and air. Suppose he had confronted the apparent clash of (a) with (b). Would he have been embarrassed? Why should he? It would have cost him nothing to concede that the arithmetical sum of the volumes of the three tetrahedra and the one octahedron would be so much smaller than that of a single icosahedron. For the question at issue, he could have pointed out, is how we are to understand space-occupancy in the two cases. In the case of the icosahedron "the room it takes up" is all contained within a single shell. But the four corpuscles into which it turns are not so contained. Nor are they stacked up against each other with no spaces in between: they are moving about the air in which they now find themselves. To do this they need a lot more space, and if deprived of this additional room by a confining vessel the impact of their aggregate collisions on its surface might well crack its walls.

A further objection might then be raised. Does not the admitted fact that the volume of the icosahedron is so much greater than the combined volumes of the three tetrahedra and the one octahedron entail that the principle of the conservation of matter has been violated? Plato would deny the entailment: his theory commits him to the conservation of matter *by surface,* not *by volume;* what it keeps constant in every transformation is the aggregate surface area of the corpuscles. If you press him to say what happens to that portion of the matter within the icosahedron which cannot be enclosed within the equivalent surface area of smaller polyhedra, Plato would say that there is no such matter: after creation matter exists only in the form of space encapsulated by polyhedra; what is not thus encapsulated is empty space, which becomes matter when captured by envelopes of the approved stereometric form. And if you retort that, even so, since his theory allows at least for conservation of space, and denies that there is space unfilled by matter,[34] Plato must still account for the space in the icosahedra that was left uncaptured by the smaller polyhedra resulting from the transformation, it is open to Plato to respond by postulating that this space will be recaptured by *other,* concurrent, transformations in the reverse direction, that is, from fire to air and from both to water, where the same aggregate area bounds a larger volume after the transformation than it did before.[35]

To return to the main issue, what is the fact here which might be thought threatening to Plato's theory? It is, in Aristotle's wording in the above quotation, that "when air is generated from water it takes up more room." Is this a fact that could "upset" Plato's theory? Surely not. So formulated it is nothing: we do not even know what it means, and cannot know until the ambiguity in the expression, "takes up more room," has been resolved. Suppose that we resolve the am-

34. That is to say, "no vacant space large enough to contain the smallest existing corpuscle of matter" (Archer-Hind, *The "Timaeus" of Plato,* p. 210). This is the only way in which Plato's declarations that "no space is left empty" (58A), "there is no void" (80C, can be understood consistently with the fact that any packing of octahedra and icosahedra is bound to leave minute interstices between adjacent solids.

35. To sustain the constancy of the total volume of the cosmos Plato would have to postulate that transformations in each direction (from *n*-sized to *m*-sized polyhedra, on one hand, and from *m*-sized to *n*-sized ones, where $n = m$) balance exactly (cf. Cornford, *Plato's Cosmology,* pp. 243–44). We might well complain that this is a huge assumption. But neither could we sustain the claim—the only one that would matter in the present discussion—that the assumption would involve any direct clash with matters of observable fact accessible to Plato at the time.

biguity by taking the expression to mean what the words are most likely to suggest: that the corpuscular constituents of air sum to a greater volume than do those of water. Then it is no longer an observable fact, and no one would claim that it is. It cannot be reckoned even a plausible inference from the observed fact that the conversion of water to steam will crack a confining container. The cracking could occur for entirely different reasons, deriving from the mobility of the discrete particles in the steam bombarding the container's walls, and having nothing to do with their aggregate volume. This might well be the reason why Plato ignores the apparent difficulty.[36] Can we fault him for doing so? Certainly we would have preferred him to face up to it and dispel the apparent inconsistency with his theory by resolving the ambiguity. But note how difficult it would have been for him to do that in the face of the fact that the volume of a gas may vary enormously, depending on the pressure, and that Boyle's law was not to be discovered until two thousand years thereafter. How can we define the alternative—and clearly relevant—sense of "takes up more room" when the space occupied by the vapor could be four cubic feet for one pressure, forty thousand cubic feet for another?

I trust I have helped explain how understandable in the circumstances is that part of Plato's conduct which is most exasperating to modern readers: his lordly insouciance for the empirical verification of his elaborate and ingenious physical theory, his indifference to confrontation with empirical fact. Only once does he refer to such a thing in the *Timaeus*, and then his language is so unguarded that it has been often misunderstood—has even been taken to say the very opposite of what it means:

> For these examples [of color-mixture] it will be tolerably clear to what mixtures the remaining colors should be likened so as to preserve the likelihood of the account. But if anyone were to put these matters to a practical test [more literally: "produce tests by

---

36. Though, of course, it need not. It is entirely possible that this whole question never occurred to Plato. See the concluding paragraph of section R in the Appendix.

investigating these matters by means of action"], he would be
ignoring the difference between man's nature and god's: god has
knowledge and power sufficient to mix the many into one, and
to resolve the one into many, but no man is equal to either task
now or ever hereafter. [68C7–D7][37]

"Why," asks a philosophical commentator on the *Timaeus*,
"does Plato allow Timaeus to shield his theory from falsification in this way?"[38] But has Plato said anything to suggest that
he anticipates *falsification*? Is he not saying that one cannot
expect any effective experimental test[39]—hence that one cannot get either verification or falsification in this way; only god
could get such a thing, and you are not god. I wonder if the
experimental scientist has ever been paid a stronger compli-
ment. Has anyone else seen the ability to construct a theory
about the world which can be actually tested in the laboratory
as a godlike power?

'But if you are going to be so humble—and so defeatist—
if you really mean that theories such as yours aren't hu-
manly verifiable, how can you expect your theorizing to
advance our knowledge?' This, I suppose, would be the
question my reader would put to Plato at this point. His

37. For a recent discussion of this passage see Lloyd, "Plato as a Natural
Scientist," p. 83.

38. I. M. Crombie, *An Examination of Plato's Philosophical Doctrines* 2:228.
Crombie is not the only one to put this reading on the passage; it has been
so read even by Shorey, "Platonism and the History of Science," p. 162, n.
2, who argues zealously for the congeniality of Plato's physical theories to
the scientific temper, and sees in this passage the one "apparent exception"
which he cannot explain away.

39. One of the reasons he was so sure of this does not seem to have been
noticed: so far as I know, no account has ever been taken of it in assessments
of the import of this passage for Plato's attitude to experimental science.
This is a consequence of the well-known fact that under the term "color"
Plato lumps properties as heterogeneous as brightness and hue (thus one of
his so-called "colors" is "the brilliant," λαμπρόν, στίλβον [67–68B]). To conduct an experimental investigation of color-mixtures he would have had to
separate out the two variables, and vary one of them while keeping the other
constant. To do this he would have had to recognize the conceptual difference between the two—which he does not: he has no concept of hue as distinct from brightness; neither does Aristotle: see R. Sorabji, "Aristotle, Mathematics, and Colour," pp. 294ff.

answer would be clear and unequivocal: "I have no such expectation." From the very start of his discourse (29Bff.), Timaeus insists that his account of the natural universe will not be knowledge—only belief; it will be no more, he says, than a "likely tale" or a "likely account."[40] What Plato is offering us he thinks *credible*—something that *could* be true: our belief in it will not bring us into collision with any hard facts; this is his modest claim. And if we were then to retort that other theories which must be false if his is true would be also credible, and ask why then we should believe his tale instead of one of these others, I think that he would only want to offer us aesthetic grounds for preferring his.[41] On these grounds, his theory would stand up well in the competition. Compare it with the best of its rivals, the Democritean. There atoms come in infinitely many sizes and in every conceivable shape, the vast majority of them being irregular, a motley multitude, totally destitute of periodicity in their de-

---

40. εἰκὼς μῦθος, εἰκὼς λόγος. There are some seventeen echoes of these key phrases in later parts of the dialogue. If we are to translate εἰκώς as "probable" (which is regularly done by Cornford and others, and is perhaps unavoidable, though "plausible" would fit some contexts better), we should bear in mind that "probable," so used, cannot bear the sense which this word normally carries nowadays in epistemological contexts, where one says that a proposition is "probable" only as a contraction for "probable *relatively to the evidence*"—evidence one is prepared to name. Plato never uses εἰκώς in this way, so far as I can recall; he could not do so without writing promissory notes which could not be cashed, for he could not produce the relevant evidence. All he can mean by saying that a physical theory of his is εἰκώς is that there is no cogent reason *against* believing it—that it is consistent with the known facts.

41. It would be hard to think of a physical theory in which aesthetic considerations have been more prominent:

> Now the question to be considered is this: What are the most beautiful (κάλλιστα) bodies that can be constructed . . . [53 D7–E1: quoted more fully on pp. 79–80]. If we can hit upon the answer to this question we have the truth concerning the generation of earth and fire . . . [E3–4] For we will concede to no one that there are visible bodies more beautiful (καλλίω) than these [E4–5]. This then we must bestir ourselves to do: construct four kinds of body of surpassing beauty (διαφέροντα κάλλει) and declare that we have reached a sufficient grasp of their nature . . . [E6–8]. Of the infinitely many [scalene triangles] we must prefer the most beautiful (τὸ κάλλιστον) . . . [54A2–3].

And so forth.

sign, incapable of fitting any simple combinatorial formula.[42]
If we were satisfied that the choice between the unordered
polymorphic infinity of Democritean atoms and the elegantly
patterned order of Plato's polyhedra was incapable of empirical adjudication and could only be settled by asking how a
divine, geometrically minded artificer would have made the
choice, would we have hesitated about the answer?[43]

In what I have said so far in this chapter, I have made no
reference to Plato's epistemological doctrine which restricts
knowledge to cognitive encounters with immaterial, super-
sensible, eternal objects, entailing that even a true physical
theory could not qualify as knowledge, but only as true belief,
since what physics purports to describe and explain is material, sensible, flux.[44] I have chosen to ignore this doctrinaire
apologia for the low epistemic ranking of the physical theories of the *Timaeus,* for this ranking makes perfectly good
sense without the apologia, indeed better sense without it

42. If my interpretation of the Epicurean *minima* is correct ("Minimal Parts
in Epicurean Atomism," pp. 145ff.), Epicurus reformed the atomic theory to
bring into it a simplicity and economy whose elegance compares favorably
with that of the Platonic theory. I would hazard the guess that the Epicurean
reform was inspired by the *Timaeus*; we know of no other physical theory
that could have served as its model.

43. But I am not suggesting that the aesthetics of the Platonic theory are
flawless. The exclusion of earth from the combinatorial scheme (necessitated
by the noninterchangeability of its cubical faces with the triangular faces of
the solids assigned to fire, air, and water) is awkward. Worse yet is the role
of the fifth regular solid, the dodecahedron, whose properties would also
shut it out of the combinatorial cycle. The hasty reference to it (55C) suggests embarrassed uncertainty. What could he mean by saying that "the god
used it for the whole"? The commentators have taken him to mean that the
Demiurge made the shape of the universe a dodecahedron; this unhappily
contradicts the firm and unambiguous doctrine of 33B (reaffirmed in 43D
and 62D) that the shape of the universe is *spherical.*

44. That only Forms qualify as objects of knowledge is Plato's doctrine in
dialogues of his middle period (*Phaedo* 65C2ff.; *Republic* 476Eff., 514Aff.).
Forms remain objects of knowledge *par excellence* in one of his latest works
(*Philebus* 58E–59C), where he concedes that "true belief" may also purvey a
kind of knowledge, though of a very inferior kind (*Philebus* 61D10–E4,
62A2–D7). In the *Timaeus* (28A1–4, 28B8–C1, 29B3–C3, 51D3–52A7) this
concession is not made. Plato is still maintaining the rigid dichotomy: knowledge (ἐπιστήμη, νοῦς) of eternal, immutable, abstract objects; belief (πίστις,
δόξα) in the case of temporal, changing, sensible things.

than with it. Consider the following statements about the liquid in our oil-lamp:

(1) It is oil.

(2) It consists of icosahedra with some tetrahedra mixed in.

Plato's formal epistemology denies the status of "knowledge" to both, demotes both to "true belief." In so doing it unhappily obscures the radical difference in their epistemic status. Plato's reasons for declining to certify (1) as knowledge, though philosophically intriguing, are scientifically irrelevant: they have no bearing on the scientist's interest in theories verifiable by hard facts. (1) is a sterling specimen of a "hard fact" in the sense I defined above: no serious investigator would have any trouble assuring himself that the stuff in the lamp is oil rather than, say, muddy water or yellowish wine; there are good practical tests, commonly agreed upon among informed persons, by which the difference between oil and these other substances may be ascertained. Not so in the case of (2): should I doubt its truth there are *no* practical tests by which I can resolve the doubt; there is nothing remotely approaching a "hard fact" in (2) itself or in any statement which Plato might conceivably have offered as the basis for (2)'s truth. So (1) and (2) must be kept distinct, while Plato's epistemology conflates them, giving us, in effect, the grounds for philosophical skepticism concerning the truth of (1) as though they were also grounds for scientific agnosticism concerning the truth of (2). This is reason enough for abstracting from Plato's general epistemological doctrine when we confront his disclaimer that (2), and all other parts of his physical theory, could constitute knowledge.[45]

A last question: Once you renounce the hope of attaining knowledge in your theories about the natural universe,

---

45. To reinforce my point: if we were to accept Plato's grounds for degrading his theory of the constitution of matter to "true belief"—because what it purports to describe and explain is sensible and temporal—then on these same grounds we would have to degrade similarly the whole of his astronomical theory; we would have to agree with him, for example, that its beautiful demonstration that the motion of the sun is a spiral must be reckoned a "true belief" and therefore cannot count as a contribution to knowledge.

would you still have good reason to engage in such theorizing? The question concerns Democritus no less than Plato,
for he too had reached darkly pessimistic conclusions concerning the possibility of achieving knowledge of nature. Two
of his fragments read:

> But in reality we know nothing;[46] for truth is in the depths. [B117]
>
> Yet it will be clear that to really know what each thing is like is
> beyond our power. [B8][47]

Yet neither he nor Plato found this terminal agnosticism de-
bilitating. They pressed forward in their natural inquiries with
unabated zest. Plato's is evident in the pages of the *Timaeus*.
As for Democritus, we have his famous saying that "he would
rather find one natural explanation than win the kingdom of
Persia" (B118).

Is this understandable? I find it so. Once you have seen
how groundless were those euphoric claims to godlike
knowledge that had been made by earlier, more naïve, *physiologoi*,[48] you still have options on what you are to believe.
So what will you do? Will you go back to the refuse heap
of superstition from which even men as intelligent as Herodotus cannot wholly tear themselves away? If not, you will
still find intellectual satisfaction in figuring out systematically what the world would be like if it were cosmos—if it
were rational through and through, every event in it inde-
feasibly governed by natural regularities. Democritus produced such a system, and it must have brought to the most

---

46. ἐτεῇ δὲ οὐδὲν ἴδμεν. In Popper, "Back to the Presocratics," p. 153,
the Greek phrase is translated, "But in fact, nothing do we know from having
seen it." The last four words are a conjectural expansion for which there is
no foundation in the wording of the Greek text (ἴδμεν is simply the Ionic
form of ἴσμεν, "we know"), nor yet in its context in the source (Diogenes
Laertius, 9.72).

47. For other Democritean fragments to the same effect see B6, B7, B9,
B10 (all from Sextus Empiricus). They are conveniently brought together in
translation in W. K. C. Guthrie, *History of Greek Philosophy* 2:458.

48. Heraclitus implicitly claims access to the wisdom "which steers all
things through all things" (B50), since that wisdom *is* the logos, "in accordance with which all things happen," which he purports to understand and
expound (B1). Both Parmenides and Empedocles present their physical doctrines in the guise of a revelation given them by a goddess.

sophisticated minds of his own age a sense of intellectual liberation such as its Epicurean version was to bring Lucretius later on. When faith in your ancestral gods has vanished so totally that nothing is left in its place, not even an aching void, and you are prepared to debate their existence as dispassionately as you would the existence of crocodiles, then Democritus has something to offer you which, if not knowledge, is the next best thing—a credo in the form of a naturalistic scenario, a coherent expression of your perception of a rational universe.

But between the fourth century B.C. and the seventeenth A.D.—throughout the two millennia that separate the death of Thucydides from the birth of Spinoza—the clientele of Democritean materialism could not be large, even among intellectuals. The vast majority found faith in the supernatural a spiritual necessity—many of them still do. For these the *Timaeus* offered a brilliant alternative. If you cannot expunge the supernatural, you can rationalize it, turning it paradoxically into the very source of the natural order, restricting its operation to a single primordial creative act which insures that the physical world would be not chaos but cosmos forever after. This Plato accomplished by vesting all supernatural power in a Creator who was informed by intelligence and was moved to create our world by his love of beauty and by his pure, unenvying, goodness. Reducing all lesser divinities to creatures, dependencies, and servants of this supreme deity; making the courses of the stars movements of rational agents performing faultlessly their chronometric tasks; grounding all physical causes in the architectonic structure impressed on matter by its divine artificer, Plato, in his own perversely original way, sustained the faith that our world is cosmos. He gave rational men a pious faith to live by in two millennia all through which science was more prophecy than reality.[49]

49. I am grateful to Professor Terry Irwin who did me the kindness to read a draft of this chapter. His queries led me to make a correction in my translation of *Ti.* 81C–D and in my phrasing in note 45.

Appendix

When the supporting material for things said in the text became too bulky and too technical, it seemed best to relegate it to this Appendix. Section A refers to p. 16.

A. In the *Iliad* (bk. 23, vv. 514ff.), when Antilochus wins the chariot race by foul means, he is challenged by Menelaus, "Swear that you did not block my chariot of your own will (ἑκών), by guile" (v. 585). This shows that he is prepared to absolve Antilochus of responsibility of the offense if he could be persuaded that the blocking of his chariot by Antilochus had been unintentional. It is Antilochus' intent, not just the act, that matters to Menelaus. I do not know how Dodds would handle either the Antilochus or the Phemius episodes, both of which are clear counterexamples to his thesis: there is no reference to either of them in his well-documented essay. Adkins, *Merit and Responsibility*, pp. 10–11, takes account of the Phemius episode and related evidence, conceding that a man *is* absolved of moral responsibility for what he does under "compulsion." But one wonders if he realizes the implications of this concession when he proceeds to speak of the "ignoring of intention" in Homeric society (p. 55) and to remark (*loc. cit.*) that Homeric society "can rarely spare a thought for intentions" (he has not argued that only rarely would duress be found acceptable as a defense). The "compulsion" on Phemius is not brute physical force: he is not dragged by the suitors, kicking and screaming, to their banquets in Odysseus' palace. He walks there himself, finding this more eligible than the alternative of exposing himself to the suitors' wrath. So it is the *intention* with which he goes (on his own legs) to their banquets that exculpates him of moral responsibility in the eyes of Odysseus and Telemachus. As for the Antilochus episode, which Adkins discusses for another purpose (p. 56), he fails to note that it is a clear counterexample to his thesis.

B. In speaking of the "wanderings" of the seven members of this group (39D1, τὰς τούτων πλάνας; 40B6-7, τὰ δὲ τρεπόμενα καὶ πλάνην τοιαύτην ἴσχοντα), Plato is acutely aware of two things: (1) that the word carries the strongest connotations of unsettled, irregular, erratic movement—motion not confined to a unique trajectory, but "straying" from one path to another; and (2) that, in his view, this implication is hopelessly wrong and misleading, since those same motions which are being called "wanderings" by others (and even by himself, when speaking with the vulgar) are a very paradigm of orderly, regular, rational movement.

For (1) cf., e.g., Plato's description of the rootless, nomadic sophists as a "wandering tribe, [moving] from one city to another, without fixed abode of its own" (γένος . . . πλανητὸν . . . κατὰ πόλεις οἰκήσεις τε ἰδίας οὐδαμῇ διῳκηκός) (Ti. 19E); and his use of the expression "the wandering cause" (τὸ τῆς πλανωμένης αἰτίας εἶδος, 48A6-7) for the domain of mechanistic causation which, operating by itself (i.e., without teleological controls), represents irrational, disorderly contingency: "separated from [teleological] intelligence they produce their sundry effects at random and without order" (after Cornford's translation). Particularly revealing for what Plato finds so offensive (even "blasphemous," Laws 821D2) in the widespread notion that sun, moon, and planets (for my use of this term, see note 19 in Chapter 2) "wander" is the remark in Laws 822A4-8: ". . . this doctrine concerning the moon and the sun and the other stars [the five planets]—that they should even wander—is incorrect. The very contrary is the case, for each of these revolves always in one and the same path—not *in many paths*-though it seems to be moving in many paths."

The error is pinpointed in the italicized phrase: this is what it means to believe in the "wandering" of these bodies. This being the case, we may notice at once that in the doctrine of lunar, solar, and planetary motion which he expounds programmatically in the *Timaeus* (discussed in the text: pp. 53ff.)—that the trajectories of all seven of these bodies are helices produced by the composition of the movements of the Same and the Different (39A-B)—Plato is *rejecting* the "incorrect" view denounced in Laws 821B-822A (instead of accepting it as has been thought by some scholars: see section D, below). The idea of the composition of motions, used by Plato in the *Timaeus* (and even in the *Republic*: cf. Cook Wilson, quoted by Adam *ad Republic* 617A), makes it possible to distinguish perfectly between moving in diverse paths (ὁδούς), on the one hand, and moving along the same path with diverse motions (πολλὰς φοράς), on the other. Thus Aristotle speaks of each of the planets moving "in its own circuit" (κατὰ τὸν αὐτοῦ κύκλον) with "multiple motions" (πλείους κινήσεις) (de Caelo 291B1-3). Cf. Heath, *Aristarchus*, p. 183, expound-

ing a point which had been made already by August Boekch in 1852
(*Untersuchungen über das kosmische System des Platons*, pp. 52ff.):
". . . the unity of the movement of the planets in single circles is not
supposed, here [in the *Laws*] any more than in the *Timaeus*, to be
upset by the fact that the movement of the circle of the Same turns
them into spirals. Thus in the *Timaeus*, in the very next sentence but
one to that about the spirals, Plato speaks of the moon as describing
'its own circle' in a month, and of the sun as describing 'its own
circle' in a year (39C3–5). Similarly, Dercyllides says that the orbits
of the planets are primarily simple and uniform circles round the
earth; the turning of these circles into spirals is merely incidental."
Thus to hold, as Plato does in the *Timaeus*, that, e.g., the sun always
moves in a spiral is to affirm in that dialogue that "it revolves always
in one and the same path—not in many paths" and hence to deny
that it "wanders" in the literal and popular sense of the term.

For (2) the all-important passage in the *Timaeus* is 39C5–D2 (with
the back-reference to it in 40B6–8), to be cited below in section D.

C. "And in order that there might be some perspicuous measure
of their relative quickness and slowness with which [reading καθ᾽
ἅ at 39B3, with Archer-Hind and others, instead of καὶ τὰ of the
codices] they moved in the eight revolutions, the god kindled a
light in the orbit which comes second from the earth, which we
now call 'sun,' that he might shine on all the heavens and that all
living beings for whom it was proper might come to possess number, taught by the revolution of the Same and uniform. Thus night-
and-day, the period of the single and wisest revolution, was generated" (39B2–C2). There is a problem here which Cornford and
most other commentators have ignored. Plato appears to be talking as though the period of the diurnal revolution of the sun—
the solar day—were *identical* with the movement of the Same, while
his theory requires it to be a little shorter: in the sun the movement of the Same is retarded, being counteracted by the *inverse*
movement of the Different. Only in the fixed stars, which are totally exempt from the latter movement, could the period of the
diurnal revolution be precisely identical with the movement of the
Same. Only A. E. Taylor, *A Commentary on Plato's "Timaeus,"* pp.
214–15 shows awareness of the difficulty, and he cuts the knot
by declaring that "the distinction between the mean solar day
and the sidereal day . . . was not yet discovered," paying no attention to the fact I have just mentioned: that on the very theory
expounded in the *Timaeus* there must be a small, but appreciable, difference between the two units. The correct solution, I surmise, is that Plato is ignoring (not denying) the difference in this
context, where he is discussing celestial periodicities in their concrete teleological function of *making living creatures aware of the fact*

that the passage of time is measurable. To qualify for this purpose the movement of the Same has to be represented by the diurnal revolution of the Sun, for this is what performs this function *in primis*: by creating the dramatic alternations of day and night it forces on our attention the recurrence of countable units of temporal passage. Moreover, for practical, calendaric, purposes, the solar day is inevitably the basic unit of measurement: calendars go by the solar, not the sidereal, day, and when Greek astronomers take a hand in the reform of the calendar by devising "Great Years" (luni-solar cycles with intercalary months) the solar day is the unit by which their months and years are measured. This, I suggest, must be the reason why Plato here (and again a few lines later, 39D6-7, τῷ τοῦ ταὐτοῦ καὶ ὁμοίως ἰόντος ἀναμετρηθέντα κύκλῳ) is content to overlook the difference between the solar and the sidereal periods of the revolution of the Same.

D. "Because few men have taken account of the periods of the rest, mankind has no names for them nor does it determine by investigation their numerical ratios; it does not know that time is so to speak *the wanderings* (πλάνας) *of these bodies,* which are so bewilderingly numerous and astonishingly variegated (πλήθει μὲν ἀμηχάνῳ χρωμένας, πεποικιλμένας δὲ θαυμαστῶς)" (39C5-D2). The phrase I have italicized (where I take the qualification, "so to speak," to apply to the clause "that time is the wanderings of these bodies"— not to the verb "to know," as in the usual translations, which makes poor sense) is most provocative. Plato is doing his utmost to bring home to the reader his conviction that these so-called "wanderings" are in fact very paradigms of regular motion, since they (no less than the sun and the moon) are meant to serve as measures of the temporal order of the universe—to be "moving image(s) of eternity" (37D5). He is telling us that in respect of regularity there is no difference between their motions and those of the sun and moon; the only difference is that the latter are so conspicuous, apparent to every last human being, while the former are known only to "a few."

From the fact that Plato in the *Timaeus* does not scruple to fall in with popular usage and speak of the "wanderings" of the planets (πλάνας here; and again in 40B6, where he refers to these and to the sun and moon as well, as τὰ τρεπόμενα καὶ πλάνην τοιαύτην ἴσχοντα), while in the *Laws* he rails against such language as false and even "blasphemous" (see section B, above), some scholars have inferred that when he wrote the *Timaeus* Plato really believed that the movements of these bodies are irregular and became convinced of their regularity only later, in the course of the interval which separates the *Timaeus* from the *Laws*. See, e.g., Owen, "The Place of the *Timaeus* in Plato's Dialogues," p. 87; Burkert, *Lore and Science,* pp. 325-26; Mit-

telstrass, *Die Rettung der Phänomene*, pp. 130ff. In section B, above,
I have already given one reason why I cannot agree with this interpretation. I may now add a second: to impute to the author of the
Timaeus the belief that those so-called "wandering" celestial motions
are irregular is to ignore the substantive doctrine of the *Timaeus*
which is affirmed so strongly in the above citation and throughout
the whole passage on the creation of time (37A6ff.). Plato could not
have held up the movements of the planets as an "image of eternity
revolving according to number" (39A7–8) unless he believed that
they instantiate periodicities of the most exact and dependable
form which can be realized in the physical universe. I say "believed,"
because nowhere in this passage does Plato profess to *understand*
or *to have learned* these "bewilderingly numerous and astonishingly variegated" movements; he only expresses confidence that
these movements *must* be regular—else they could not constitute
"time"—though all too few "have taken account" (or "considered,"
"reflected upon") their periods: the word Plato uses here, ἐννενοηκότες, should be taken in its primary sense (see Liddell and Scott,
Greek-English Lexicon, s.v. ἐννοέω, I.2), and cf. Archer-Hind's translation of ἐννενοηκότες ("have not taken into account") and
Apelt's ("haben . . . keine Aufmerksamkeit geschenkt"); mistrans-
lations (Bury, "discovered"; Cornford, "observed") would yield
spurious textual evidence for thinking that Plato is saying here that
the length of all the planetary periods (including Saturn's thirty-
year period) have already been determined (cf. n. 74 in Chap. 2,
above).

E. The two accounts which have long been recognized as closest
to the original Theophrastean account of Anaximander's system are
Hippolytus, *Refutatio Omnium Haeresium* [hereafter to be cited by
author's name only] 1.6.5. [= DK 12A11] and pseudo-Plutarch,
Strom. 2 [= DK 12A10]. Neither of these mentions planets or says
anything to imply that they were informed that such bodies figured
in Anaximander's world model. A place for planets in Hippolytus has
been made solely by the conjecture that the words καὶ τῶν πλανήτων
had been originally present in the text after ἀπλανῶν in the phrase
τοὺς τῶν ἀπλανῶν κύκλους. The conjecture is textually arbitrary and
is motivated by the assumption that Hippolytus would not have written ἀπλανῶν except in contrast to "wandering" stars *other than the
moon and the sun:* the error of the assumption is manifest once one
stops to think that for Theophrastus (or indeed for every Greek writer
on astronomy) ἀπλανῆ could be used in contrast to sun and moon
(cf. Chap. 2, n. 19, above) no less than to the five planets; and that
this *is* the intended contrast in the passage is clear enough in its
reference to the circles of the sun and the moon, followed by

ἀνωτάτω μὲν εἶναι τὸν ἥλιον in the second clause, whose μὲν is answered by the δὲ in the final clause, κατωτάτω δὲ τοὺς τῶν ἀπλανῶν ἀστέρων. The only mention of planets in Anaximandrian doxography occurs in Aetius 2.15.6 [in DK 12A18]—a patently confused text which ascribes the same order of sun, moon, fixed-stars-and-planets to an incongruous trio consisting of Anaximander, Metrodorus (fourth-century atomist), and Crates (fourth-third-century cynic). As for Simplicius (*in de Caelo* 471.2ff. [= DK 12A19]), the introductory phrase, περὶ τῆς τάξεως τῶν πλανωμένων καὶ περὶ μελεθῶν καὶ ἀποστημάτων, is followed up by the report of Eudemus (fr. 146 Wehrli) that "the Pythagoreans [in all probability Philolaus: see Chap. 2, n. 65, above] were the first to discover the positional order (τὴν τῆς θέσεως τάξιν)" of the "wandering" stars. Since Anaximander could hardly have made a place for the planets in his rigidly structured architectonic scheme without fixing their position in it relative to the sun, the moon, and the fixed stars, this testimonium must be counted as definite evidence *against* the presence of planets in the Anaximandrian model.

F. Burkert debunks effectively the ascription to Pythagoras of the identity of Morning and Evening Stars (*Lore and Science*, pp. 307–8). But Burkert does concede the ascription of the distinction between fixed stars and planets to two other predecessors of Parmenides: Anaximenes and Alcmaeon (pp. 310ff., 332ff.). I submit that, though widely accepted, in neither case will the ascription stand up to critical scrutiny. In the case of Anaximenes not a single text in our sources credits him, directly or by implication, with any opinion which implies the existence of planets. The ascription has been reached on the strength of the supposition that the conflict between two texts,

(1) . . . similarly the sun and the moon and all the other stars, being fiery, float on air (Hippolytus, 1.7.6 [in DK 41A14]),
(2) Anaximenes holds that the stars are fastened like nails in the crystalline [sphere] (Aetius 2.14.3 [in DK 41A14]),

must be resolved by understanding Hippolytus to mean "planets" when writing "stars"! The simpler explanation, surely, is that Hippolytus is unaware of the theory ascribed to Anaximenes in (2), while the veracity of (2) would be suspect even apart from this conflict, both because of its content (a crystalline sphere is improbable in sixth-century astronomical theory, and is scarcely reconcilable with Anaximenes' known physical theory: see Guthrie, *History of Greek Philosophy* 2:134–35) and because it implicitly contradicts another text in Aetius' (frequently thoughtless) compilation, namely 2.23.1 [in DK 13A15] which agrees perfectly with Hippolytus' testimony in (1).

As to Alcmaeon, the whole case for his having discovered the existence of "the planets" (which ones? all of them?) rests on the following doxographic text: "Some of the mathematicians hold that the
planets move from west to east, contrary to the fixed stars. Alcmaeon too agrees with these men" (Aetius 2.16.2-3 [in DK 24A4]).
Now who are "the mathematicians" to whom the first period refers?
If they were Pythagorean near-contemporaries of Alcmaeon as has
sometimes been assumed (most recently by Guthrie; *History of Greek
Philosophy* 1:357), this would lend some plausibility to the doxogra-
pher's afterthought about Alcmaeon in the second period. But this
assumption is baseless. As Burkert points out (*Lore and Science*, pp.
42-43, n. 76; his documentation could be strengthened with additional texts) the μαθηματικοί of the doxographers are the pro-
fessional mathematicians and, most particularly, mathematical astronomers of the Hellenistic age (a usage which may derive from
Aristotle's occasional use of μαθηματικοί for "mathematical astronomers," as in *de Motu Anim.* 639B7 and *Metaph.* 1073B11-12). In
their case—that of post-Aristotelian mathematical astronomers—the
eastward movement of the planets is *de rigueur*. The extension of
that doctrine to Alcmaeon in the above text (probably by incorpora-
tion of somebody's marginal note) has no support in any other doxographic text and none in Aristotle's testimony: when enumerating
(exhaustively: τὰ θεῖα πάντα) the heavenly bodies in Alcmaeon's
theory, Aristotle mentions only "moon, sun, stars" (*de Anima*
405A29-B1). The dubiousness of the extension is shown up by the
following consideration (cf. Burnet, *Early Greek Philosophy*, p. 110,
n. 2): the eastward motion of the planets is a piece of sophisticated
astronomy (ascribed to Oenopides [Chap. 2, n. 42, above], discov-
erer of the obliquity of the ecliptic and of the Great Year of 59 years
[Chap. 2, n. 43, above]), while Alcmaeon's astronomy was primitive:
its sun is "flat," its moon is "bowl-shaped" and its eclipses are
caused by tiltings of the bowl (Aetius 2.22.4 and 2.29.3 [in DK
24A4]), which shows (among other things) that he was unaware of
the fact that the moon is not self-luminous, which was discovered by
Parmenides (B14 and B15; other texts and commentary in O'Brien,
"Derived Light and Eclipses," pp. 118ff).

G. The information supplied in our sources does not state, or
definitely imply, that Democritus did know of the existence of the
five planets. All we get is, first, that one of his treatises was entitled
"About the planets" (B5f); then a couple of doxographic reports,
one of them (Aetius 2.15.3 [DK 68A86]) informing us that Democritus recognized Venus as a planet, the other (Hippolytus, 1.13.4 [in
DK 68A40]) that he thought the planets were "not all at the same
height," i.e., at the same distance from the earth. From the fact that

he wrote a treatise on the planets we may safely infer that he devoted serious investigations to the topic; and from the title of one of his other treatises, ἐκπετάσματα, "Planispheres," i.e., presumably plane projections of the celestial sphere (cf. Ptolemy, *Geog.* 7.7), we may infer that he constructed celestial maps. That he must have known the five planets (known to his contemporary, Philolaus) would be, at best, a reasonable presumption.

Burkert (*Lore and Science*, p. 313) holds that we can learn more from the following in Seneca (*Naturales Quaestiones*, 7.3.2 [68A92]: "Democritus quoque, subtilissimus antiquorum omnium, [A] suspicari se ait plures stellas esse quae currant, sed nec numerum illarum posuit nec nomina, [B] nondum comprehensis quinque siderum cursibus." Assuming that "plures" in [A] has its primary sense (*plus* = "more"), Burkert (followed by Guthrie, *History of Greek Philosophy* 2:420) takes Seneca to be saying that Democritus suspected that there were five *more* (i.e., over and above the canonical quintet), thereby implying that Democritus did know the latter. I cannot follow this acrobatic reasoning. The metonymic use of "plures" to mean "a great number, many" is common in Latin prose (see Lewis, *Latin Dictionary*, s.v. *plus*). And this is most likely to be the sense here (as has been generally assumed, most recently by G. J. Toomer, *Oxford Classical Dictionary*, 2d ed. [1970], *s.v.* "Astronomy," p. 134), for two reasons: In the first place, an implied back-reference to five planets is excluded (Seneca had made no allusion to them in the antecedents of the quotation). In the second place, Seneca could hardly have meant to offer [B] in explanation of Democritus' "suspicion" that there were more planets than the five (knowing of the existence of five planets while not knowing their trajectories could hardly give one a reason for suspecting that there was an indeterminately large number of planets), while [B] could very well have been offered in explanation of Democritus' "suspicion" that there was an indeterminately large number of planets (one who does not know of either the existence or the trajectories of the five planets could very well suspect that a "very large number" of planets is required to explain the phenomena). So if we had reason to think that Seneca was well informed on the content of Democritus' book about the planets, we would have to take the citation from him as evidence that Democritus did *not* know the quintet. But the overwhelming probability is that Seneca was going by what he had learned from an intermediate source (which may have echoed the Aristotelian report of Democritus' theory of comets: cf. Chap. 2, n. 62, above), inferring that Democritus believed in "a very large number" of planets from that theory as reported and criticized in Aristotle's *Meteorologica*, and hence only *assuming* that Democritus did not know the canonical five.

H. The term Plato uses in *Ti.* 40C5 is ἐπανακυκλήσεις (cf. ἐπανα-κυκλούμενον, *Republic* 617B, of the behavior of Mars). The force of ἀνα- in Plato's use of ἀνακύκλησις is clear from *Politicus* 269E, where the noun is used to describe the *reversal* of the cosmic revolution related in the myth, and thus means a *counter-revolution* (translated "revolution in reverse" by J. B. Skemp, *Plato's "Statesman"*; "retrogradation circulaire" by A. Diès, *Platon, "Le Politique"*). The force of ἐπι- and ἀνα- in ἐπανακύκλησις = "a counter-revolution that comes *on* a revolution": in the course of revolving in a given direction a body starts revolving in the reverse direction. This would well explain Plato's use of the phrase τὰς τῶν κύκλων πρὸς ἑαυτούς ἐπανακυκλήσεις (*Ti.* 40C5) to express the notion that a revolution in a given direction has occasional set-backs during which it loops backward. Proclus, whose text read ἀνακυκλήσεις, takes it for granted that the term means "retrogradations" (προποδισμούς) in this passage (*Comm. in Timaeum*, 248C).

If we give this sense to ἐπανακυκλήσεις in *Ti.* 40C5, as is done generally by the commentators, we can hardly avoid doing the same to ἐπανακυκλούμενον in *Republic* 617B (so, e.g., L. Robin, *Platon, Oeuvres Complètes*, who translates "animé . . . d'un mouvement retro-grade"), which leaves us wondering why Plato should have mentioned the retrogradation of Mars without a word about that of the other planets (which he could hardly have ignored if he knew that of Mars and which he had no motive to leave unmentioned when mentioning that of Mars). The simplest solution seems to me to suppose with Burnet (*Early Greek Philosophy*, p. 304, n. 1) that the received text is defective and should be expanded by adding after ἐπα-νακυκλούμενον the phrase μάλιστα τῶν ἄλλων which occurs in Theon of Smyrna's quotation of the passage. As had been previously observed (Dreyer, *History of Astronomy*, p. 59, n. 3; Duhem, *Le Système du Monde* 1:60, 110) it makes very good astronomical sense to speak of Mars "retrograding the most among the planets" (its arc of retrogradation is the greatest). Burnet gives two good reasons for endorsing the restoration, in spite of its complete lack of authority in our own codices: (1) Theon "is apparently quoting from Dercyllides, who first established the text of Plato from which ours is derived"; (2) "μάλιστα τῶν ἄλλων is exactly fifteen letters, the normal length of omissions in the Platonic text."

I. A further reference to retrogradation I see in a notoriously cryptic phrase (38D3–6): Venus and Mercury were made "to revolve in a course which is on a par with the sun's in respect of speed, but *they received the power contrary to his* (τὴν δὲ ἐναντίαν εἰληχότας αὐτῷ δύναμιν) as a result of which (ὅθεν) they overtake, and are overtaken by, one another alike. . . ." The italicized phrase is a commentator's bugbear. To run through the various interpretations and assess their

comparative merit would call for a paper of no small length (for
sample discussions see Heath, *Aristarchus of Samos*, pp. 165–69, and
A. E. Taylor, *Commentary on Plato's "Timaeus,"* pp. 196–202.) Van
der Waerden's subsequent revival of the view of some of the ancient
commentators that Plato is here postulating epicycles ("Die Astronomie der Pythagoreer," pp. 53ff.) fails to disarm the *prima facie* anachronism of such a view. As von Fritz points out, "The epicyclic theory
explains the appearances much better than does the theory of homocentric spheres of Eudoxos of Cnidos; . . . hence it would be
difficult to account for the origin of the latter if the epicyclic theory
had already been discovered at an earlier time" (*Grundprobleme der
Geschichte der antiken Wissenschaft*, p. 174; he proceeds to anticipate
and rebut van der Waerden's answer to this grave—surely, decisive—
objection).

The most natural reading of τὴν δὲ ἐναντίαν εἰληχότας αὐτῷ δύναμιν would be "they received the direction contrary to his." This
is how T. H. Martin (*Études sur le Timée* 2:70ff.) had insisted that
Plato's phrase should in fact be read, and his view has been treated
with surprising tolerance in the subsequent literature (most recently
by Dicks, *Early Greek Astronomy*, p. 124; but cf. ibid., p. 126): if accepted, this view would have devastating consequences, making
Plato so much of an astronomical ignoramus as to render his views
in this area not even worth discussing. To take Plato to be saying that
the annual motion of Venus and Mercury is contrary to the sun's
annual motion would be to give him a theory of planetary motion "in
evident contradiction with the most easily observed facts" (Martin,
Études sur le Timée, p. 72). Now if τὴν δὲ ἐναντίαν αὐτῷ δύναμιν had
stood alone, Martin's interpretation would have great force, in spite
of its horrendous consequences (though even then one would have
wondered why, if this were what Plato wanted to say, he did not say
it more simply and directly, e.g., by writing, ἐκείνῳ δὲ ἐναντίον
κύκλον ἰόντας). But the fact is that the puzzling phrase is followed
immediately by a sentence which states a *consequence* (ὅθεν); and this—
that the sun and the two planets "overtake, and are overtaken by,
each other"—is unintelligible as a consequence of the supposition that
both of the two planets are moving in a direction contrary to the
sun's. For this reason Martin's view must be ruled out by the clear
implications of Plato's text (for further arguments against it see Taylor, *Commentary on Plato's "Timaeus,"* pp. 196ff.).

If, on the other hand, we do take the phrase as a reference to
retrogradation—as we may be understanding τὴν ἐναντίαν αὐτῷ
δύναμιν to mean "the power [to move in ways] contrary to his
own"—the consequence follows easily, since in the course of getting
ahead of, or falling behind, one another their predominantly eastward motion would yield intermittently to stretches of westward motion, and in these stretches they would be moving contrary to the

sun whose own motion was uniformly eastward. This is Proclus' interpretation (*in Timaeum* 221Dff., E. Diehl, ed. pp. 67–68), and Taylor (p. 201) argues that it has a tradition behind it which goes back to the first generation of the Academy. To the question, 'If retrogradation is what Plato has in view here, why does he speak only of Venus and Mercury? Why not also of the three outer planets, to one of which (Mars) he had ascribed it directly (ἐπανακυκλούμενον) in the *Republic*?,' the answer is, surely, that he is explicitly denying us here an account of the "others" (i.e., of Mars, Jupiter, and Saturn), saying that this he might *perhaps* (ἴσως) do adequately (ἀξίως) at some future time (38D6–E2): declining to go into the particulars of the motions and periods of the outer planets (to no one of which does he make so much as a single individual reference), he has no occasion to speak of their retrogradation. I have italicized the "perhaps": Plato is not professing to have the requisite knowledge and withholding it from the reader; he is expressing the hope of being able to do at some future time the job that he is in no position to do now.

J. Maria Cardini, "Sui passi controversi di Platone," pp. 33–35, has argued convincingly that the doctrine of *Ti.* 39A5–B1 is anticipated, more tersely and less clearly, in 36D4–5, κατὰ τἀναντία μὲν ἀλλήλοις ἰέναι προσέταξεν τοὺς κύκλους, if we read this difficult sentence as follows: τἀναντία must be taken as a substantive (grammatically possible, though its alternative, adverbial, use is much more frequent), and ἀλλήλοις as a neuter referring to τἀναντία (not as a masculine referring to κύκλους). So read, the sentence says that *each of the seven circuits is progressing in two contrary directions (at the same time)*, i.e., has a complex movement resulting from the two contrary motions of the Same and the Different. The construction is difficult, but not impossible, and I believe that it should be accepted: it makes excellent sense, and moreover, as Cardini points out, its correctness is supported by the recurrence of κατὰ τἀναντία ἰέναι a little later, in 39A6–B, διχῇ κατὰ τὰ ἐναντία ἅμα προϊέναι. (For the reference to Cardini I am indebted to Harold Cherniss.) An alternative attempt to get the same sense out of 36D4–5, while adhering to the traditional construction of the grammar (τἀναντία adverbial and ἀλλήλοις masculine), is to take ἀλλήλοις to refer to the κύκλοι of the Same and the Other. Dicks (*Early Greek Astronomy*, pp. 128–29) refers to this as "possible (though perhaps unlikely)," and points out that it commended itself to Proclus (*Comm. in Timaeum* 221F). So far as I can see the proposed reference of ἀλλήλοις is *not* a possible one in that context: if it did refer to "circuits," the only ones to which it would refer are the "seven" which had been mentioned just before (ἑπτὰ κύκλους) in 36D2.

K. Nowhere in the *Timaeus* is there a reference to the "third force" postulated by Cornford, nor yet in *Laws* 898D–E, which he adduces (*Plato's Cosmology*, p. 108) as additional evidence: *the most* one can get out of that passage is that the stars are animated self-moving beings, which we already know from the *Timaeus*; I emphasize "the most" because it is not clear that even that much can be extracted from *Laws* 898D–E: at D3–4 we are not told that their own individual souls drive "the sun and the moon and the other stars" around the heavens; all we are told is that "soul" (ψυχή) does so; that the soul-force is that of individual souls "indwelling" in each star is only one of the three possibilities laid out in the curious sequel at 898E5–899A4. What is not said or implied here or in the *Timaeus* is that the self-motion of the souls of the moon and of the five planets *accounts for those movements of theirs which are unaccountable by the motions of the Same and the Different,* which is the very crux of Cornford's claim. It would be well to note that the self-motion of the moon and the planets would cover *whatever motions they happen to have* and would be properly exercised in those movements of theirs which *are* accountable by the composition of the motions of the Same and the Different. Though, certainly, it was the Demiurge who originally "bestowed upon" (προσῆψεν, 47A7) each celestial soul these two motions, they become for each thereafter its very own, perpetually self-initiated, motions. In his "Table of Celestial Motions" (p. 136) Cornford himself (implicitly) recognizes this in the case of the World Soul; he (very properly) lists the Same and the Different as the World Soul's own "self-motions" in spite of the fact that they are imparted to it at its creation by the Demiurge (who "shared out" [ἀπένειμεν, 34A4] circular motion to it and "made it revolve" [ἐποίησε κύκλῳ κινεῖσ-θαι, 34A4]). There is, therefore, no reason to assume—as Cornford appears to do in various parts of his exposition, and conspicuously in his "Table of Motions"—that the self-motions of the stars have to be motions other than those imparted to them by the Same and the Different. So far as Plato's astral theology and psychology are concerned, all the planets would be fully endowed with the power of self-motion even if they had no other motions but these (except of course, for axial rotation as well, which is not in controversy).

A further objection to Cornford's proposal is that it leaves totally unexplained the extraordinary obscurity of the phrasing to which Plato resorts in his account of the motion of Venus and Mercury in 38D3–6 (cf. section I above). If Plato had really held that it was the exercise of these two planets' power of self-motion which made them speed up or slow down the motion of the Different, he would have had no difficulty in saying so. Why then should he, instead, invoke, so cryptically, their possession of "the contrary power"? I can think of no explanation other than the embarrassment in which

he finds himself at this point. The excuse he gives a little later—that
a fuller description "would be futile without visible models" (40D2-
3)—is hardly convincing: even without visual aids a writer of Plato's
expository powers could have done more for the reader than toss
him that strange phrase, "[Mercury and Venus] received the power
contrary to his [the sun's]" (38D4), which would be astronomical
nonsense if given the obvious, natural, reading.

Von Fritz too (*Grundprobleme der Geschichte der antiken Wissenschaft*,
pp. 172 and 175), insists on the significance of the obscurity of expression in the *Timaeus* when Plato's account abuts on discrepancies
between his astronomical theory and the empirical phenomena; von
Fritz rightly sees here a difficulty for van der Waerden's ascription
to Plato of an epicyclic theory (see section I, above): were the ascription true, it would have made Plato's resort to these dark sayings
unnecessary.

L. Our judgment of the historical reliability of Sosigenes' report
that Plato had challenged astronomers to produce a hypothesis of
uniform and regular motions that would "save" the apparently irregular motions of the planets depends entirely on what we make of
lines 18–24 on p. 488 of Simplicius' commentary on the *de Caelo*. I
had best quote the passage in full, including its part (b) (quoted in
translation in Chap. 2 above and in Greek in the accompanying note
95, so that its relation to the preceding part (a) may be judged in
context):

> And (a) Eudoxus of Cnidus was the first Greek to engage in such hypothe-
> ses [*sc.* of orderly circular motions saving the phenomena of planetary
> motion], as Eudemus records in Book II of his History of Astronomy,
> and Sosigenes too, having got this from Eudemus. For (b), as Sosigenes
> says, Plato had set this problem to those who were engaged in these stud-
> ies: What uniform and orderly motions must be hypothesized to save the
> phenomenal motions of the stars?

From the fact that Simplicius starts off by stating in (a) that
Eudemus was Sosigenes' source for the report about Eudoxus, and
then proceeds in (b) to quote Sosigenes' report about Plato without
saying that here too Sosigenes' source is Eudemus, it has been inferred (most recently by J. Mittelstrass, *Rettung der Phänomene*, p.
154, n. 214; cf. von Fritz, *Grundprobleme der Geschichte der antiken
Wissenschaft*, p. 179, n. 375, referring to Mittelstrass and agreeing
with him on this point) that what is reported about Plato in (b) was
not in Eudemus. To underwrite this inference categorically we
would have to exclude two possibilities: (1) Sosigenes may have
failed to acknowledge his indebtedness to Eudemus in his report
about Plato in (b). (2) The dependence on Eudemus may have been
quite clear in Sosigenes' text and Simplicius may have failed to ac-

knowledge it for purely stylistic reasons—because of a disinclination to reiterate the reference to Eudemus as Sosigenes' source for (b) which he had made for (a) in the preceding period. I fail to see cogent grounds for concluding that both (1) and (2) must be ruled out. In the case of (1) we cannot be sure that Sosigenes would cite Eudemian authority for each of the statements in his book deriving directly or by inference from Eudemus' *History of Astronomy.* So far as phrasing goes there is nothing in the quotation from Sosigenes in (b) which could not have echoed Eudemus. In particular, the crucial phrase "to save the phenomena" could represent the diction of either Sosigenes or Eudemus: it occurs in a passage (*in de Caelo* 497.15-24 [= Eudemus fr. 149 Wehrli; I quoted it in part in n. 34 above in Chap. 2]) where Simplicius appears to be quoting Eudemus directly: λέγειν γὰρ αὐτόν φησιν (sc. Eudemus) appears to introduce a quotation; and Simplicius' qualitative description of the Eudemian report (συντόμως καὶ σαφῶς ὁ Εὔδημος ἱστόρησεν) is good evidence that he is referring to the Eudemian account itself rather than to a paraphrase of it in Sosigenes. As for (2), since in (b) Simplicius shifts from report to citation, he has to indicate this fact, and for this purpose a phrase like ὥς φησι Σωσιγένης is indispensable. So the fact that Sosigenes is mentioned twice is not itself the serious matter it has been often thought to be. The real question is rather whether or not Simplicius' failure to reiterate in (b) the availability of Eudemian authority is due to his wish to minimize verbal repetitiveness or to convey the positive message that the Eudemian authority which exists for (a) simply does not exist for (b). It would take no small faith in the historiographic fastidiousness of Simplicius to be sure of the second alternative. I conclude that both possibilities (1) and (2) remain open. But since that is the most that can be said for either of them, no historical argument could be grounded on the assumption which older historians had not scrupled to make (e.g., Duhem, *Sōzein ta phainomena,* p. 3; Heath, *Aristarchus of Samos,* p. 140) that Sosigenes' report about Plato in (b) represents historical fact.

M. The phrase "saving the phenomena" does not occur in the Platonic corpus nor yet in Aristotle's works. In Plato "save a thesis (or 'argument')" (e.g., *Tht.* 164A, εἰ σώσοιμεν τὸν λόγον) or "save a tale" (*Laws* 645B, ὁ μῦθος ἀρετῆς σεσωσμένος ἂν εἴη) and in Aristotle "save a hypothesis" (*de Caelo* 306A30, σῴζειν τὴν ὑπόθεσιν) and "preserve a thesis" (*Nic. Eth.* 1096A2, θέσιν διαφυλάττων) occur in contexts where "to save" is to preserve the credibility of a statement by demonstrating its consistency with apparently recalcitrant logical or empirical considerations. The phrase "saving the phenomena" must have been coined to express the same credibility-salvaging operation in a case where phenomena, not a theory or an argument, are being put on the defensive and have to be rehabilitated by a rational account which resolves the *prima facie* contradictions besetting their uncritical acceptance. This is a characteristically Platonic view of phenomena. For Plato the phenomenal world, symbolized by the shadow-world in the Allegory of the Cave (*Rep.* 517B1–3) is full of snares for the intellect. Thus, at the simplest level of reflection, Plato refers us (*Rep.* 602C7ff.) to illusions of sense, like the stick that looks bent when partly immersed in water, or the large object that looks tiny at a distance. Thrown into turmoil (ταραχή, C12) by the contradictory data of sense, the soul seeks a remedy in operations like "measuring, numbering, weighing" (D6) so that it will no longer be at the mercy of the phenomenon (ὥστε μὴ ἄρχειν ἐν ἡμῖν τὸ φαινόμενον, D7–8). For Plato, then, the phenomena must be held suspect unless they can be proved innocent ("saved") by rational judgment. So it would not be surprising if the phrase "saving the phenomena"—showing that certain perceptual data *are* intelligible after all— had originated in the Academy, though we have no means of proving that it did. The phrase would have described aptly, for example, what Plato does for the optical data which are accounted for by his theory of vision (*Ti.* 46A2–C6; *Sph.* 266B6–C4): the phenomena of the reversal of left and right in ordinary mirrors and of the reversal of up and down in cylindrical mirrors of vertical curvature would be intellectually disturbing if taken at face value for they would then clash with our normal perceptual beliefs; Plato's theory puts him in a position to show that such appearances must, nonetheless, "of necessity" be what they are (πάντα τὰ τοιαῦτα ἐξ ἀνάγκης ἐμφαίνεται, 46B1); he could thus have said with perfect justice that his theory "saves" those phenomena.

N. From the examples in the text above it can be easily seen that a great variety of air-fire compounds could result from a given body of water under different physical conditions. All four examples have been predicated on the assumption that the physical conditions to which a given body of water is subject are such as to cause all of it to change in accordance with a single formula, all of it changing either to fire, as in (a), or to air, as in (d), or to a ($3F$, $1A$) mixture, as in (b), or to a ($1F$, $2A$) mixture as in (c). Under suitably different conditions the resultant fire-air mixtures could be very different. Thus suppose (counterfactually) that there were conditions in which half of the water changed according to the formula in (b), the other half according to the formula in (c). We would then get

(e) $2W \longleftrightarrow (4F + 3A)$ $[= (3F + 1A) + (1F + 2A)]$

Here we would get a mixture much warmer than in (c), but less hot than in (b). This shows how, given suitable conditions, air with a maximal or minimal fire-content could result from the same body of water. Thus, given 10,001 corpuscles of water, we could get under appropriately different conditions,

$$10,001W \longleftrightarrow (1F + 25,002A),$$

at one extreme—one corpuscle of the water changing according to formula (c), all the rest according to formula (d)—to yield $[(1F + 2A) + 25,000A]$, or

$$10,001W \longleftrightarrow (50,003F + 1A),$$

at the other extreme—one corpuscle of the water changing according to formula (b), all the rest according to formula (a)—to yield $[(3F + 1A) + 50,000F]$, or any one of a vast number of intermediate varieties, ranging from the lowest to the highest temperatures which air could reach according to Plato's physical theory.

O. I am assuming that Plato is using here the phrase ἐπίκλην καλούμενος in the same way as he does ἐπίκλην ἔχοντα of the planets in 38C (where it is clear that he is *not* endorsing the view on which a commonly used term had been predicated: cf. above, sections B and D of this Appendix). By so reading the phrase we can avoid a difficulty which has been ignored by the commentators (it is partly concealed in some translations by the use of different words to render ὁμίχλη in different passages: Cornford translates it "murk" in 58D, "mist" in 66E; Moreau, in the Bude translation, edited by L. Robin, translates "brouillard" in 58D, "buée" in 66E). If Plato were declaring that "mist and darkness" are forms of air, he would be contradicting the plain implications of other statements of his in the *Timaeus*. His account of vision (45B–D) gives no quarter to the quaint Greek notion that makes of "darkness" a special stuff to fill the atmosphere after sundown; in Plato's theory this darkness represents only the loss of sunlight caused by the sinking of the sun below the horizon, *not* the influx (from where?) of dark-colored gas. As to "mist," Plato states unambiguously at 66E that it is "air on the way to water" (τὸ ἐξ ἀέρος εἰς ὕδωρ ἰόν). Plato could hardly have used this phrase if he had committed himself a few pages earlier to the doctrine that mist is a variety of air. Elsewhere (*Platonic Studies*, pp. 366ff.) I have argued that by "air on the way to water" Plato is most likely to refer to transitional states through which a mass of air would pass in the course of changing into water, the air-component steadily decreasing as air-corpuscles keep turning into water corpuscles (I would now add: probably only to states at advanced stages of this process, where the water component is already appreciable and the mixture is noticeably moist).

In the case of "aether," by saying that there is a variety of air which is (popularly) so called, Plato could be putting off a decision on what place this controversial element should have in his own cosmology. Empedocles had identified aether with air (B71, 2; B98, 2; B109, 2), while Anaxagoras had contrasted it sharply with air (B2; B15), making this contrast the primordial opposition from which all other differentiations evolve in his cosmogony. In the *Phaedo* Plato refers to the aether as that which "most of those who discourse about such things [i.e., most of the *physiologoi*] call 'the heavens'" (109B–C). In the myth of that dialogue (a highly imaginative, but not wholly fanciful, composition) aether stands to air as air does to water, hence (*pace* A. Rivaud, trans, and ed., *Timée, Critias*, p. 24) aether is a perfectly distinct element. Xenocrates in his *Life of Plato* ascribes to Plato "five figures or bodies, namely aether, fire, water, earth, air" (*ap.* Simplicius, *in Phys.* 1165, 33ff.). This suggests that by the end of his life Plato came around to a view which he had expressed tentatively in the myth of the *Phaedo*, that the aether is a fifth element on a par with the other four—which is the firm doctrine of the *Epinomis* (881E) and is made the physical basis of Aristotle's astronomical theory in the *de Caelo*. If Plato were still undecided on this point when composing the *Timaeus*, he would not wish to say that aether was a special variety of air; quite understandably then, he would refer to the latter noncommittally as ἐπίκλην "αἰθὴρ" καλούμενος.

P. The explanation of the workings of the oil lamp that might be offered by the Democritean theory, as suggested briefly in the text above, might raise further questions in the reader's mind. For example:

1. How is it that oil may be very cool while containing a full quota of fire-atoms, and that its temperature may vary from ice-cold to boiling hot while the fire-water ratio in its composition remains constant?

2. Why do the fire-atoms in the oil keep streaming into the burning wick?

I do not think that these questions would have stumped Democritus any more than those I have raised in the text above. He has a principle (reported casually by Theophrastus, *de Sensibus* 65 [= Democritus A135]) that "what contains most void grows hottest" (evidently, because density would dampen down atomic mobility), which would enable him to answer (1) (claiming that the density of the oil decreases as it gets warmer, so that the same quota of fire-atoms will comport, at appropriately different rates of agitation, with the lowest as with the highest temperatures of the oil) and even (2) (arguing that the concentration of fire-atoms moving at random in the oil will tend to be heaviest in the part of the liquid which is closest to the burning wick, where the oil is warmest and thus "contains the most void").

Still another question might be raised:

3. If water-atoms are escaping upward, how is it that no vapor is ever visible, as happens in the parallel case of evaporating water, and that no water is recoverable from the air into which these water-atoms allegedly move?

There is no need to suppose that this would embarrass Democritus any more than would (1) and (2): all he would need to say is that the two cases are *not* parallel; the water-atoms in the oil belong to a special variety which, when liberated into the air, does not behave as does ordinary water-vapor. It is not visible while dispersing into air nor recoverable from it by the familiar processes of condensation (though it might be by other means, presently unknown).

But the reader would get a distorted picture of ancient science if he were led to think by the above discussion that in the science of the time questions such as the ones I have scouted were systematically anticipated and answered along some such lines as those I have sketched. Had that been the case both the conceptual texture of physical theory and the range of its application would have been vastly different from what we must presume it to have been on the basis of the surviving fragments and those essays (like the *Timaeus*, the physical treatises of Aristotle, and the poem of Lucretius) which have survived intact. What we in fact get is a set of imaginative guesses which fit well enough the selected set of salient phenomena that were at the forefront of discussion, guesses which are so conveniently flexible (so ingenious and so vague) that they will stretch to cover any further phenomena that ordinary observation could present.

Bibliography

Adam, J. *The "Republic" of Plato*. 2 vols. Cambridge: Cambridge University Press, 1902.

Adkins, A. W. H. *Merit and Responsibility*. Oxford: The Clarendon Press, 1960.

———. *From the Many to the One*. Ithaca, N.Y.: Cornell University Press, 1970.

———. *Moral Values and Political Behaviour in Ancient Greece*. London: Chatto and Windus, 1972.

Apelt, O. *Platons Dialoge "Timaios" und "Kritias."* Leipzig: Meiner, 1919.

Archer-Hind, R. D. *The "Timaeus" of Plato*. London: Macmillan, 1888.

Bicknell, P. "Early Greek Knowledge of the Planets," *Eranos* 68 (1970): 47–54.

Blass, Fr., ed. *Eudoxi "Ars Astronomica" qualis in charta Aegyptiaca superest*. Kiel, 1887.

Bloom, A. *The "Republic" of Plato*. New York: Basic Books, 1968.

Bruins, E. M., "La Chimie du Timée," *Révue de Metaphysique et Morale* 56 (1951): 269ff.

Burkert, W. *Lore and Science in Ancient Pythagoreanism*. Trans. E. L. Minar, Jr. Cambridge, Mass.: Harvard University Press, 1972.

Burnet, John. *Early Greek Philosophy*. 4th ed. London: A. and C. Black, 1945.

Bury, R. G., trans, and ed. *Plato: Timaeus, Critias, Cleitophon, Menexenus, Epistles*. Cambridge, Mass.: Harvard University Press, 1929.

Cardini, M. T. "Sui passi controversi di Platone, *Timeo* 40B, 36C, 36D; *Leges* 822AC. Di Aristotele, *de Caelo* 293A15–293B33," *La Parola del Passato* 40 (1955):20–40.

Cherniss, H. *Aristotle's Criticism of Plato and the Academy*. Vol. 1. Baltimore, Md.: The Johns Hopkins Press, 1944.

Cornford, F. M. *Plato's Cosmology: The "Timaeus" of Plato*. London: Routledge, 1937.

———. "Was the Ionian Philosophy Scientific?" *Journal of Hellenic Studies* 62 (1942): 1ff.

———. *The "Republic" of Plato.* New York: Oxford University Press, 1945.

———. *Principium Sapientiae.* Cambridge: Cambridge University Press, 1952.

Crombie, I. M. *An Examination of Plato's Philosophical Doctrines.* Vol. 2. London: Routledge, 1963.

Cumont, F. "Les noms des planètes et l'astrolatrie chez les Grecs," *Antiquité Classique* 4 (1935): 5–43.

Denniston, J. D., and D. Page. *Aeschylus' "Agamemnon."* Oxford: The Clarendon Press, 1957.

Dicks, D. R. "Solstices, Equinoxes, and the Presocratics," *Journal of Hellenic Studies* 86 (1966): 26–40.

———. "On Anaximander's Figures," *Journal of Hellenic Studies* 89 (1969): 120.

———. *Early Greek Astronomy to Aristotle.* London: Thames and Hudson, 1970.

Diels, H., and W. Kranz. *Die Fragmente der Vorsokratiker.* 3 vols. 6th ed. Berlin: Weidmann, 1951–52.

Diès, A. *Platon, "Le Politique"* (vol. 9, part 1, in *Platon, Oeuvres Complètes*). 2d ed. Paris: Les Belles Lettres, 1950.

Dijksterhuis, E. J. *The Mechanization of the World Picture.* Trans. C. Dikshoorn. Oxford: The Clarendon Press, 1961.

Dodds, E. R. *The Greeks and the Irrational.* Berkeley: University of California Press, 1951.

Dryer, J. L. E. *A History of Astronomy from Thales to Kepler.* 2d ed., revised with a Foreword by J. E. Stahl. New York: Dover, 1953. (Reprint of Dreyer's classical treatise, *History of the Planetary Systems from Thales to Kepler.* Cambridge: Cambridge University Press, 1906.)

Duhem, P. *Sōzein ta phainomena.* Paris: Hermann, 1908.

———. *Le Système du Monde.* Vol. 1. Paris: Hermann, 1913.

Düring, I. *Aristoteles.* Heidelberg: Winter, 1968.

Festugière, A. J. "Platon et l'Orient," *Révue de Philologie* 21 (1947): 5–45.

Friedländer, P. *Plato.* Vol. 1. Trans. H. Meyerhoff. Princeton, N.J.: Princeton University Press, 1958.

Furley, D. J., and R. E. Allen, eds. *Studies in Presocratic Philosophy.* Vol. 1. London: Routledge, 1970.

Gigon, O. *Der Ursprung der griechischen Philosophie von Hesiod bis Parmenides.* Basel: Schwabe, 1945.

Guthrie, W. K. C. *A History of Greek Philosophy.* Vol. 2. Cambridge: Cambridge University Press, 1965.

Hammond, N. G. L., and H. H. Scullard. *Oxford Classical Dictionary.* 2d ed. Oxford: The Clarendon Press, 1970.

Heath, T. *Aristarchus of Samos, the Ancient Copernicus*. Oxford: The Clarendon Press, 1913.

Heidel, W. A. "*Peri Physeōs*," *Proceedings of the American Academy of Arts and Sciences*, Boston (1910), pp. 77–133.

Holwerda, D. *Physis*. Groningen: Wolters, 1955.

Kahn, Charles. *Anaximander and the Origins of Greek Cosmology*. New York: Columbia University Press, 1960.

————. "On Early Greek Astronomy," *Journal of Hellenic Studies* 90 (1970), 99–16.

Kerschensteiner, Jula. "*Kosmos*," *Zetemata*, Heft 30. Munich, 1962.

Keyt, D. "The Mad Craftsman of the *Timaeus*," *Philosophical Review* 80 (1971): 230–35.

Kirk, G. S. *Heraclitus: The Cosmic Fragments*. Cambridge: Cambridge University Press, 1954.

————. and J. Raven. *The Presocratic Philosophers*. Cambridge: Cambridge University Press, 1957.

Koyré, A. *Metaphysics and Measurement: Essays in the Scientific Revolution*. London: Chapman and Hall, 1968.

Kuhn, T. S. *The Copernican Revolution*. Cambridge, Mass.: Harvard University Press, 1957.

Laserre, F. *The Birth of Mathematics in the Age of Plato*. London: Hutchinson, 1964.

————. *Die Fragmente des Eudoxos von Knidos*. Berlin: de Gruyter, 1966.

Lee, H. D. P., trans. and ed. *Plato: "Timaeus."* Baltimore: Penguin Books, 1965.

Lesky, A. "Göttliche und menschliche Motivation im homerischen Epos," *Sitzungsberiche der Heidelberger Akademie der Wissenschaften, Philosophische-historische Klasse*, 1961 (4. Abhandlung).

Lewis, Ch. T. *Latin Dictionary for Schools*. Oxford: The Clarendon Press, 1889.

Liddell, H. G., and R. Scott. *Greek-English Lexicon*. New ed. by H. S. Jones. Oxford: The Clarendon Press, 1925.

Lindsay, A. D., trans. and ed. *The "Republic" of Plato*. London: Dent, 1935.

Lloyd, G. E. R. "Plato as a Natural Scientist," *Journal of Hellenic Studies* 88 (1968):78–92.

Lloyd-Jones, H. *The Justice of Zeus*. Berkeley: University of California Press, 1971.

Marcovich, M. *Heraclitus: Greek Text with Short Commentary*. Merida, Venezuela, 1967.

Martin, T. H. *Études sur le Timée de Platon*. 2 vols. Paris: Ladrange 1841.

Mittelstrass, J. *Die Rettung der Phänomene*. Berlin: de Gruyter, 1962.

Morrow, G. R. *Plato's Cretan City*. Princeton, N.J.: Princeton University Press, 1960.

Nestlé, W. *Vom Mythos zum Logos.* Stuttgart: Kröner, 1942.

Neugebauer, O. *The Exact Sciences in Antiquity.* 2d ed. Princeton, N.J.: Princeton University Press, 1957.

O'Brien, D. "Derived Light and Eclipses in the Fifth Century," *Journal of Hellenic Studies* 88 (1968): 114–27.

————. "Anaximander and Dr. Dicks," *Journal of Hellenic Studies* 90 (1970): 198.

Owen, G. E. L. "The Place of the *Timaeus* in Plato's Dialogues," *Classical Quarterly*, n.s. 3 (1953): 79–95. (Reprinted in *Studies in Plato's Metaphysics*, ed. R. E. Allen, pp. 313–38.)

Pannekoek, A. *A History of Astronomy.* English translation of *De Groei van ons Werelbeeld* (Amsterdam, 1951); name of translator not given. New York: Rowman, 1961.

Pohle, W. "The Mathematical Foundations of Plato's Atomic Physics," *Isis*, vol. 62 (1971).

Popper, Karl. "Back to the Presocratics," *Conjectures and Refutations.* 3d ed. London: Routledge, 1969. (Reprinted in *Studies in Presocratic Philosophy*, ed. D. Furley and R. E. Allen.)

Ranulf, S. *The Jealousy of the Gods and Criminal Law at Athens.* London: Williams and Norgate, 1933.

Rivaud, A. *Platon, "Timée," "Critias."* (Vol. 10 in *Platon, Oeuvres Complètes*.) 3d ed. Paris: Les Belles Lettres, 1956.

Robin, L., trans, and ed. *Platon: Oeuvres Complètes.* 2 vols. Paris: Pléiade, 1950.

Ross, W. D. *Aristotle's "Metaphysics."* Vols. 1 and 2. Oxford: The Clarendon Press, 1924.

Samburski, S. *The Physical World of the Greeks.* Trans. M. Dagut. London: Routledge, 1956.

Shorey, P. "Platonism and the History of Science," *Transactions of the American Philosophical Society* 66 (1927): 159–82.

————. *Plato: The "Republic."* 2 vols. London: Heinemann. Vol. 1, 1930. Vol. 2, 1935.

Skemp, J. B. *Plato's "Statesman."* London: Routledge, 1952.

Sorabji, R. "Aristotle, Mathematics, and Colour," *Classical Quarterly* 22 (1972): 293–308.

Tannery, P. *Pour L'Histoire de la Science Hellène.* 2d ed. Paris: Gauthiers-Villars, 1930.

Tarán, L. *Parmenides.* Princeton, N.J.: Princeton University Press, 1965.

Taylor, A. E. *A Commentary on Plato's "Timaeus."* Oxford: The Clarendon Press, 1928.

van der Waerden, B. L. "Die Astronomie der Pythagoreer," *Verhandlungen der koninklijke Nederlands Akademie van Wetenschappen*, Afd. Natuurk. I (1951), Amsterdam, 53ff.

————. *Science Awakening.* Trans. A. Dresden. Groningen: Noordhoff, 1954.

Vlastos, G. "Slavery in Plato's Thought," *Philosophical Review* 50 (1941): 289–304. (Reprinted in Vlastos, *Platonic Studies*, pp. 147–63.)

———. "Equality and Justice in Early Greek Cosmologies," *Classical Philology* 42 (1947): 156–78. (Reprinted in D. J. Furley and R. E. Allen, eds., *Studies in Presocratic Philosophy* 1:56–91).

———. "The Physical Theory of Anaxagoras," *Philosophical Review*, vol. 59 (1950). (Reprinted in A. D. P. Mourelatos, ed., *The Pre-Socratics*.)

———. "On Heraclitus," *American Journal of Philology* 76 (1955): 337ff.

———. Review of Cornford's *Principium Sapientiae* in *Gnomon*, 27 (1955): 65–76. (Reprinted in *Studies in Presocratic Philosophy*, ed. D. J. Furley and R. E. Allen.)

———. "Creation in the *Timaeus*: Is it a Fiction?" In *Studies in Plato's Metaphysics*, ed. R. E. Allen, pp. 401–19.

———. "The Disorderly Motion in the *Timaeus*." *Classical Quarterly* 33 (1939): 71–83. (Reprinted in *Studies in Plato's Metaphysics*, ed. R. E. Allen, pp. 379–99).

———. "Minimal Parts in Epicurean Atomism," *Isis*, vol. 56 (1965).

———. *Platonic Studies*. Princeton, N.J.: Princeton University Press, 1973.

Von Fritz, K. *Philosophie und sprachlicher Ausdruck bei Demokrit, Plato, und Aristoteles*. New York: Steckert, 1938.

———. *Grundprobleme der Geschichte der antiken Wissenschaft*. Berlin: de Gruyter, 1971.

Index of Greek Words

ἀήρ, 87
αἰθήρ, 114
αἰτία, 99
αἴτιος, 16
ἀναίτιος, 16
ἀνακύκλησις, 106
ἀπονέμω, 109
ἀπλανής, 102, 103
ἀρχή, 68
ἀστήρ, 103
ἄτη, 13-17, 22

δημιουργός, 26
διαθιγή, 67
διάζωσις, 39
δόξα, 94

εἰκώς, 93
ἐκπέτασμα, 105
ἑκών, 16, 98
ἕλιξ, 55
ἐναντίος, 55, 107, 108
ἐναργής, 35
ἐννοέω, 102
ἐπανακυκλέω, 106, 108
ἐπανακύκλησις, 50, 106
ἐπίκλη, 32, 113, 114
ἐπισημασία, 37
ἐπιστήμη, 94
Ἕσπερος, 43, 44
Ἑῷος, 44
Ἑωσφόρος, 43, 44

θάτερον, 34

θεῖος, 104
θέσις, 111

ἰσάριθμος, 36
ἰσόδρομος, 49

κατὰ διάμετρον, 34
καταντικρύ, 50
κατὰ πλευράν, 34
κίνησις, 60, 99
κοσμέω, 3
κόσμος, xii, 3, 4, 6, 7, 18
κύκλος, 49-51, 99, 101, 102, 106-9

λαμπρόν, 92
λόγος, 8, 26, 49, 96
λόξωσις, 39

μαθηματικός, 104
μεσόνυξ, 43
μετέωρος, 20
μέτρον, 35
μῦθος, 111

νοῦς, 26, 94

ὁδός, 99
ὁλκή, 77
ὀμίχλη, 72, 113
ὁμόδρομος, 49
ὁμοίωσις θεῷ, 28
ὄργανα χρόνων, 35, 49
οὐρανός, 36

123

παραβολή, 50
πίστις, 94
πλάγιος, 34
πλάνη, 33, 99, 101
πλανήτης, 43, 99, 102
πλανητόν, 32
πλανώμενος, 60, 99, 103
προποδισμός, 106
προχώρησις, 50

ῥυσμός, 67

σκότος, 72
σπέρμα, 67
στεφάνη, 44
στίλβον, 92
στοιχεῖον, 67, 68, 69
σύμφασις, 45
σύναψις, 50
σώζειν, 111

τάξις, 103

τρέπω, 32, 99, 101
τροπή, 32, 67

ὕβρις, 16
ὕδωρ, 113
ὑπόθεσις, 111
ὑποτίθημι, 60

φαινόμενον, 60, 112
φάσμα, 11
φθόνος, 27
φιλανθρωπία, 27
φορά, 99
φυσιολογία, 9, 11, 21, 30, 40, 42,
    47, 49, 62, 66
φυσιολόγος, 9–13, 17–19, 23–25,
    29, 30, 48, 64, 66, 80, 81, 84,
    86, 87, 96, 114
φύσις, 18–22

χορεία, 50

ψυχή, 109

Index of Names

Adam, J., 99
Adkins, A. W. H., 15, 98
Aeschines, 13, 15
Aeschylus, 18, 27
Alcmaeon, 41, 47, 103, 104
Allen, R. E., xxiv, 7, 21, 47, 69, 87
Anaxagoras, 20, 25, 38, 40, 45, 47,
 61, 64, 67, 68, 81, 82, 114
Anaximander, xxiv, 6, 10, 20, 21, 29,
 30, 39, 40-45, 102, 103
Anaximenes, xxiv, 6, 21, 80, 103
Apelt, O., 32, 80, 102
Apollonius, 65
Archer-Hind, R. D., 80, 90, 100, 102
Aristophanes, 18, 27

Bicknell, P., 43
Blass, F., 37
Boekch, A., 100
Bruins, E. M., 89
Burkert, W., 32, 46, 54, 101, 103-5
Burnet, J., xxiii, 104, 106
Bury, R. G., 80, 102

Callippus, 37, 39
Camus, A., 22
Cardini, M. T., 108
Cherniss, H., 32, 108
Cornford, F. M, xxiv, 32, 47, 48, 58,
 59, 69, 73, 80, 84, 87, 90, 93, 99,
 100, 102, 109, 113
Crates, 103
Crombie, I. M., 92
Cumont, F., 43

Democritus, xxiv, xxv, 20, 22, 38, 42,

45, 46, 47, 62, 63, 64, 67, 70, 82,
 83, 85, 86, 88, 93, 94, 96, 97, 104,
 105, 114, 115
Denniston, J. D., 27
Dercyllides, 39, 100, 106
Dicks, D. R., 32, 37, 38, 39, 41, 42,
 43, 44, 47, 49, 54, 59, 107, 108
Diels, H., 4, 18, 39, 59
Diès, A., 23, 32, 106
Dijksterhuis, E. J., 54
Diodorus Siculus, 39
Diogenes Laertius, 41, 44, 96
Diogenes of Apollonia, 25
Dodds, E. R., 13, 14, 15, 16, 27, 98
Dreyer, J. L. E., 42, 106
Duhem, P., 54, 55, 60, 106, 111

Empedocles, 20, 47, 67, 68, 81, 82,
 87, 88, 96, 114
Epicurus, 94, 97
Euctemon, 37, 38, 39, 40, 85
Eudemus, 33, 37, 39, 46, 60, 103,
 110, 111
Eudoxus, 37, 49, 54, 57, 60, 61, 65,
 107, 110
Euripides, 27

Favorinus, 41, 43
Friedländer, P., 72, 74-76, 78
Furley, D. J., xxiv, 7, 21, 47, 87

Galileo, xxv, 62
Gigon, O. A., 44
Guthrie, W. K. C, 5, 44, 96, 103-5

Heath, T. L., 32, 37, 38, 44, 46, 47,
 56, 99, 107, 111

Heidel, W. A., 18
Heisenberg, W., xxv
Heraclitus, 4–10, 12, 20–22, 25, 29, 47, 82, 96
Herodotus, 10, 11, 18, 19, 27, 41, 66, 67, 96
Hesiod, 18, 34, 43
Hipparchus, 39, 54, 65
Hippocrates, 18
Hippolytus, 64, 102–4
Holwerda, D., 18
Homer, 4, 13, 18, 27, 34, 43

Irwin, T., 97

Kahn, C, 3, 39, 41, 42
Keyt, D., 29
Kirk, G. S., 5, 6, 8, 21, 41
Kerschensteiner, J., 5, 8
Koyré, A., 62
Kranz, W., 4, 18
Kuhn, T. S., 46, 60, 65

Laserre, F., 65
Lee, H. D. P., 80
Lesky, A., 16
Leucippus, xxiv, 20, 21, 30, 47, 67
Lewis, C. T., 105
Liddell, H. G., 102
Lloyd, G. E. R., xxv, 87, 92
Lloyd-Jones, H., 16, 27
Lucretius, 64, 97, 115
Lycurgus, 14

Mahoney, M. S., 65
Marcovich, M., 5, 6
Martin, T. H., 55, 107
Meton, 37, 38, 39, 40, 85
Metrodorus, 103
Mittelstrass, J., 101–2, 110
Moreau, T., 113
Morrow, G. R., 23, 39

Nestlé, W., 11
Neugebauer, O., 37, 38, 45
Nicias, 17, 18

O'Brien, D., 41, 47, 104
Oenopides, 39, 40, 104
Owen, G. E. L., 101

Page, D., 27

Pannekoek, A., 43
Parmenides, 20, 43, 44, 45, 47, 68, 96, 103, 104
Phaenus, 37
Philip of Opus, 42
Philolaus, 46, 103, 105
Pindar, 11, 12, 66
Pohle, W., 69
Popper, K., 65, 96
Proclus, 39, 40, 50, 106–8
Ptolemy, 33, 37, 39, 54, 65, 105
Pythagoras, 29, 43, 46, 54, 103, 104

Ranulf, S., 27
Raven, J. E., 21, 41
Rivaud, A., 80, 114
Robin, L., 106, 113

Samburksy, S., xxiii
Sappho, 18
Scott, R., 102
Seneca, 105
Sextus Empiricus, 96
Shorey, P., 92
Simplicius, 37, 49, 57, 59, 60, 80, 103, 110, 111, 114
Skemp, J. B., 106
Socrates, 61
Sophocles, 14, 15, 18
Sorabji, R., 92
Sosigenes, 59, 110, 111
Spinoza, B., 97
Stesichorus, 43
Stocks, T. L., 81

Tannery, P., 41
Tarán, L., 44
Taylor, A. E., 45, 100, 107, 108
Thales, xxiv, 6
Theognis, 4
Theon of Smyrna, 33, 39, 106
Theophrastus, 37, 80, 102, 114
Thucydides, 17, 18, 22, 97
Toomer, G. J., 105

van der Waerden, B. L., 64, 107
Von Fritz, K., 39, 63, 67, 107, 110

Wilson, J. C., 99

Xenocrates, 114
Xerxes, 11

Index of Passages

Aeschines
  Against Ctesiphon 111   13
Aeschylus
  Agamemnon 757–62   27
  Prometheus Vinctus 9–11   27, 28
Aetius
  de Placitis Philos. Reliquiae
    2.7.1   45
    2.7.7   46
    2.12.3   39
    2.14.3   103
    2.15.3   104
    2.15.6   103
    2.15.7   44
    2.16.2–3   104
    2.22.4   104
    2.23.1   103
    2.29.3   104
    3.11.3   46
Anaxagoras
  B2   114
  B4   67
  B12   25
  B13   25
  B15   114
Anaximander
  B1   21
  12A5   39
  12A22   39
Aristophanes
  Plutus 86ff.   27, 28
Aristotle
  de Anima 405A29–B1   104
  de Caelo
    270A5–16   42

    291B1–3   99
    292A3–4   42
    293A20–28   46
    294B13–14   38
    297B23–30   47
    298A3–6   39
    305B11–15   88
    306A30   111
    360A1–7   81
  de Generatione et Corruptione
    315B28ff.   67
    325B15ff.   67
  de Motu Anim. 639B7   104
  Metaph. 985B13–19   67
    986A6–12   46
    1073B11–12   104
    1073B32ff.   37
  Meteorologica
    342B27–29   45
    343B10–11   42
    343B27–28   45
    343B28–30   42
  Nic. Eth.
    1096A2   111
    1107A9ff.   28

Democritus
  A135   114
  B5f   104
  B5i   67
  B6   96
  B7   96
  B8   96
  B9   96

B10	96	Hippolytus	
B117	96	*Refutatio Omnium Haeresium*	
Diodorus Siculus		1.6.5	102
Bibliotheca historica 1.98.2	39	1.6.8	64
Diogenes Laertius		1.6.9	64
Vitae Philosophorum		1.7.6	103
2.2	41	1.13.4	104
8.83	41	Homer	
9.23	44	*Iliad*	
9.72	96	19.86–89	16
		22.199–200	8
Empedocles		23.514ff.	98
B71, 2	114	23.585	98
B98, 2	114	*Odyssey*	
B109, 2	114	15.404	34
Eudemus (ed. Wehrli)		22.351	16
Fr. 145	33, 39	22.356	16
Fr. 146	103		
Fr. 148	60	Leucippus	
Fr. 149	37, 111	B2	30
		Lucretius	
Heraclitus		*de Rerum Natura*	
B1	96	5.622–34	64
B2	8	5.627–43	64
B8	7	5.644–49	64
B30	4, 8	Lycurgus	
B41	22	*Against Leocrates* 92	14
B50	96		
B73	8	Parmenides	
B76	5	B10, 5	44
B80	7	B14	104
B89	8	B15	104
B94	9	Pindar	
B119	22	Ninth Paean	11–12
Herodotus		Plato	
2.24.1	11	*Epinomis*	
2.45.3	19	881E	114
2.53.3	11	987A	42
2.68.1	19	987B	49
2.71	19	Epistle VII 324A6	47
2.109.3	41	*Laches* 194Dff.	18
7.37.2	11	*Laws*	
8.13	11	645B	111
8.129	11	760D	34
Hesiod		821B–822A	99
Theogony		821B9	32
381–83	43	821C	44
Works and Days		821D2	99
479	34	822A–C	64
564	34	822A4–8	99
663	34		

822A8–C5	49	*Timaeus*	
885B	61	19E	99
885Bff.	23	28A1–4	94
888C	23	28B8–C1	94
889B1–C6	23	29A1	52
889B4–5	23	29Bff.	93
898A3–6	51, 52	29B3–C3	94
898D–E	109	29E	27
898D–899D	51	29E–47D	28
898D3–4	109	29E–47E	66
898E5–899A4	109	30B	29
898E7–899A4	51	30Cff.	25
899A7–9	51	31A–B	29
Phaedo		33B	94
65B	53	33B7	29
65C2ff.	94	34A	31, 52
65E6–66A6	53	34A4	109
66A	52	35Aff.	31
78C6	52	36C	34
97Cff.	30	36Cff.	51
97C–99C	25	36D2	108
97D–E	30	36D4–5	55, 108
109B–C	114	36D5–7	49
Phaedrus 245C–E	31	36E2–3	32
Philebus		37A6ff.	102
56C3–4	52	37B5	31
58E–59C	94	37B6	35
61D10–E4	94	37D5	29, 101
62A2–D7	94	38B6	35, 36
Politicus 269D	52	38C	113
Republic		38C–39D	35
379Bff.	51	38C5–6	32
476Eff.	94	38D	45
479A1–3	52	38D1–3	49
479E7–8	52	38D3–6	106, 109
514 Aff.	94	38D4	110
517B1–3	112	38D4–6	58
602C7ff.	112	38D6–E2	108
602C12	112	39A	34, 64
602D6	112	39A–B	54, 99
602D7–8	112	39A2–3	49
616B–617D	47	39A5–B1	55, 108
616E–617B	45	39A7–8	102, 108
617A	99	39B2–C2	100
617A–B	49	39B3	100
617A5	34	39C3–4	35
617A8–B1	49	39C3–5	100
617B	106	39C5–D	35, 100, 101
Theaetetus 164A	111	39D1	99
Sophist 266B6–C4	112	39D2–7	34, 49

39D6-7	101
40A-B	35
40A7-B1	35
40B	32
40B1	35
40B6	32, 101
40B6-7	99
40B6-8	100
40C4-D3	50
40C5	106
40C9-D2	62
40D2-3	110
41E	49
41E5	35
42D	35
43A	52
43D	94
45B-D	113
46A2-C6	112
46B1	112
47A7	109
47E-69B	66
47Eff.	28
48A6-7	99
48B-C	68
49B7-C8	80
51D3-52A7	94
53B1-5	70
53D-E	79
53D7-E1	93
53E3-8	93
54A2-3	93
54D-E	73
54D6	67
55A8	67
55C	94
56B	67
56D-E	86
56D-57A	77
56E	86
57A-C	79
57C7-D5	73
58A	90
58D	72, 73, 113
58Dff.	83
60A	86
60A-B	71
60D3-4	84
60D4-E2	84
62C-63E	86
62D	94

66E	113
67C-68D	73
67E-68B	92
68C7-D7	92
80C	90
81C-D	97
83B-C	77
[Plutarch]	
Strom. 2	102
Proclus	
Comm. in Timaeum	
221Dff.	108
221F	108
248C	106
284C	50
Comm. in Euclid. (ed.	
Friedlein)	40
Ptolemy	
Geog. 7.7	105
Phaseis	37
Syntaxis Mathematica	
3.1	37
9.2	65
Seneca	
Naturales Quaestiones 7.3.2	105
Simplicius	
Comm. in Aristot. de Caelo	
471.2ff.	103
488.18-24	110
488.21-24	60
493.11-497.5	50
493.16-17	58
497.15-24	111
497.17-22	37
Comm. in Aristot. Phys.	
24, 26	80
1165, 33ff.	114
Sophocles	
Antigone 621-24	14
Theognis	
677-78	4
Theo Smyrnaeus (ed. Hiller)	
Expos. rerum mathem.	
128	33
191-92	39
Theophrastus	
de Sensibus 65	114
de Sign. 4	37
Thucydides	
History 7.50.4	17